Beautiful Experiments

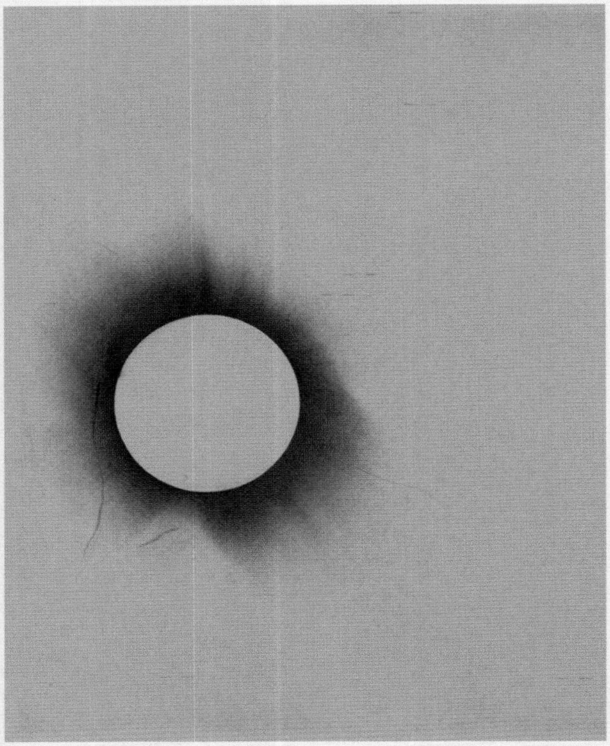

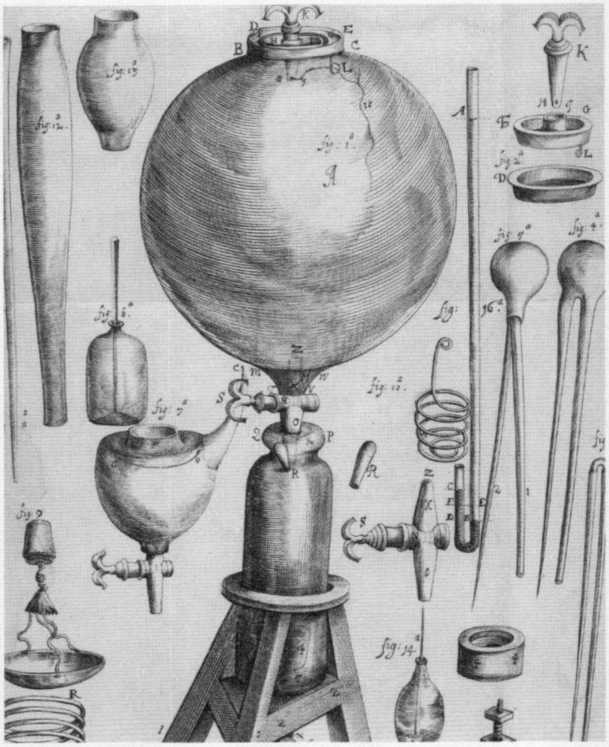

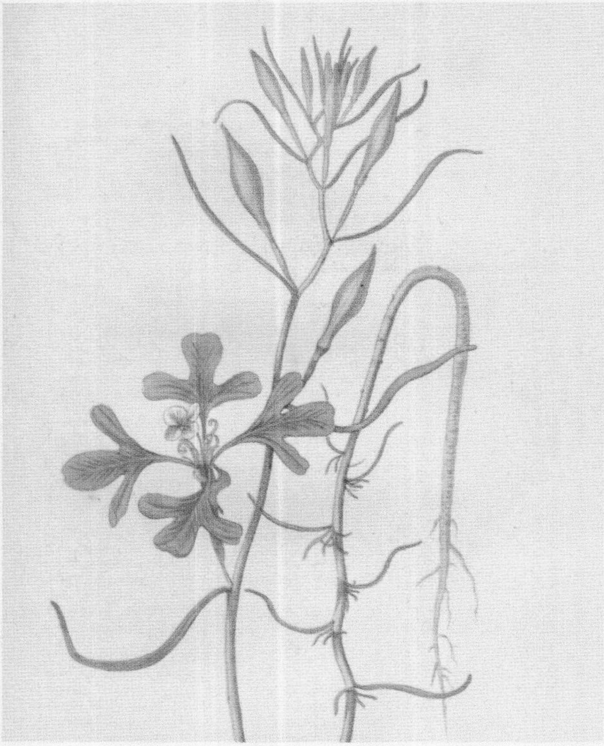

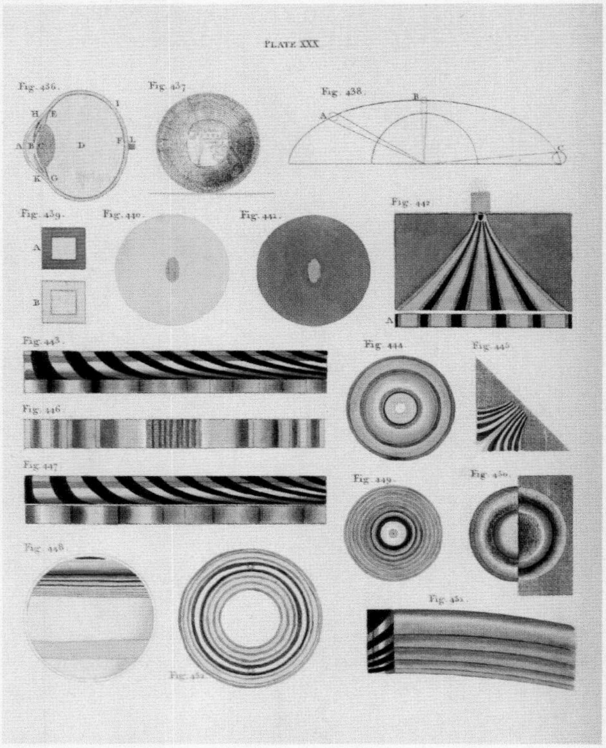
PLATE XXX
Fig. 436.
Fig. 437
Fig. 438.
Fig. 439.
Fig. 440.
Fig. 441.
Fig. 442
Fig. 443.
Fig. 444.
Fig. 445
Fig. 446.
Fig. 447
Fig. 448
Fig. 449.
Fig. 450.
Fig. 451.
Fig. 452

Beautiful Experiments

An Illustrated History of Experimental Science

Philip Ball

The University of Chicago Press
Chicago and London

The University of Chicago Press, Chicago 60637
The University of Chicago Press, Ltd., London

Published 2023

Printed in China

32 31 30 29 28 27 26 25 24 23   1 2 3 4 5

ISBN-13: 978-0-226-82582-3 (cloth)
ISBN-13: 978-0-226-83026-1 (e-book)

DOI: https://doi.org/10.7208/chicago/9780226830261.001.0001

Conceived, designed and produced by
Quarto Publishing plc
1 Triptych Place
London SE1 9SH
www.quarto.com

QUAR.355840

Library of Congress Control Number: 2023930047

Front cover: Biblioteca Nazionale Centrale, Florence (foreground); science equipment illustrations © Vector Hut / Creative Market (background)

Previous page (clockwise from top left): Natural History Museum Library, London; Science History Institute, Philadelphia; Wellcome Collection, London (PDM); Missouri Botanical Garden, Peter H. Raven Library, via BHL

Contents

A NOTE ON THE BOOK'S STRUCTURE

The chapters are thematic, while their internal organization is chronological. Chapters 3 and 4 are subdivided for clarity, and their subsections have short introductions to the key themes.

Introduction: The Experimental Philosophy

Experiments are at the core of science. It is typically by experimentation that scientific discoveries are made—from, say, the discovery of the first virus in 1892 to the creation of the vaccines against the Covid virus SARS-CoV-2 in 2020. We encounter it at an early age, from school experiments with weights and springs or the squeaky pop of hydrogen ignited in a test tube. We might be tempted to assume, then, that the process by which experiments lead to reliable and useful knowledge is understood. But that's not really so. One of the aims of this (highly selective) history of experimental science is to show how it has not been a steady accumulation of knowledge by turning the well-oiled wheels of science's experimental methodology, but something altogether more haphazard, contingent, and also more interesting and ingenious.

A history of this kind, which focuses on specific important and often elegant experiments, is necessarily constrained by the fact that some of the most significant experiments (in the broadest sense) in the history of humankind lie forgotten before recorded history: as when, for example, probably sometime in the second millennium BC in the Middle East, someone first discovered that heating iron ore with charcoal in a furnace releases the molten metal, and thereby ushered in the Iron Age. Countless ancient medicines were found by experimental trial and error—including many that were doubtless useless (or worse) but also some with genuine therapeutic benefits. It's often suggested that such discoveries hardly count as experiments, as they were the results of happy accidents rather than systematic manipulations of nature's materials. This is surely unfair. Ancient craftspeople often used quite precise recipes to make commodities such as paints, dyes, glass, cement, and cosmetics, which must have been honed by careful observation. There is every reason to suppose that they would have been actively seeking novelty: maybe changing this ingredient for that will make a brighter color, a more effective cure?

Artisanal work was long neglected in the history of science—an oversight that is being corrected today, but which surely reflects a perceived intellectual hierarchy that places theories foremost. Experimentation that might be practically useful has long been regarded as a low-status activity: it was manual work, not philosophy. As British biologist Peter Medawar put it, "applied science" was vulgar, while pure science without any applied goal or product was "laudably useless." In this view, the purpose of experiments is to advance theories: to create new knowledge of the world, not merely some new artifact. But this, too, is untrue. Plenty of scientific experimentation today, especially in chemistry and materials science, aims to make a useful or perhaps simply an interesting new substance. Our material circumstances are very much the better for this kind of experiment.

All the same, the experimental laboratory *is* the foundry of new understanding too. (That metaphor is figurative; there is a whole history to be told about the variety of workplaces for experimental science, from makeshift workshops to the high-tech industrial labs of the modern era.) The philosophies of the ancient world, from Babylonia to Greece to China, were not—as is sometimes implied—by any means devoid of investigative methods that we might today be happy to call experimentation. The treatise on optics by the second-century Egyptian Roman philosopher Ptolemy, for example, describes an experiment in which a coin inside a cup viewed at an angle where

Ernest Board's twentieth-century oil on canvas of Roger Bacon in his observatory at Merton College, Oxford, Wellcome Collection, London.

it is just hidden by the rim comes into view when water is poured in, demonstrating the bending of light rays by refraction. Ancient Greek writers describe experiments on hydraulics, air, and water pressure. Aristotle says that animal dissections disprove the idea that the sex of an embryo is determined by which side of the womb it develops in; the later Greek physician Galen made extensive use of animal vivisection to understand anatomy. Greek texts are full of assertions along the lines of: "if you do X, you'll see Y." It's likely that some of these assertions were not put to the test (some are, indeed, manifestly absurd), but they show that the Greek philosophers valued *experience* and not just abstract reasoning.

In Classical and medieval texts, the words *experientia* and *experimentum* may be more or less synonymous. Thus, "experiments" in the Middle Ages were often conducted to demonstrate rather than to evaluate a theory or an idea: experience confirms them to be true. The seventeenth-century English philosopher Francis Bacon, often considered the father of the "experimental philosophy," drew an important distinction between knowledge acquired by random chance—experience presents it to you—and by deliberate action. Only the latter, he said, is a true experiment.

Yet experiments cannot be just a bunch of random stuff that we observe. How, then, do you make them more than that? Bacon sought to formulate an answer in his book *Novum Organum* (1620), which explained how observations could be systematically accrued so that one might progress from specific facts to general axioms. His method was complicated and never really put to use by the "experimental philosophers" he inspired. But more importantly, he made the case for experimentation as the best way to understand the world, defending it against the accusation that, because it involved artificial manipulations, perhaps demanding fancy instruments, it couldn't possibly pronounce on how unadulterated nature behaves. This was one of the objections to the newly invented microscope: that it didn't just magnify but distorted the image of the specimen. Bacon countered that "artificial things differ from natural things not in form or essence, but only in the efficient" (that is, how they were produced).

Light refracted by a spherical glass full of water, according to Ptolemy. From Roger Bacon's *De multiplicatione specierum* (1275–1300), Manuscript Royal 7 F VIII, The British Library, London.

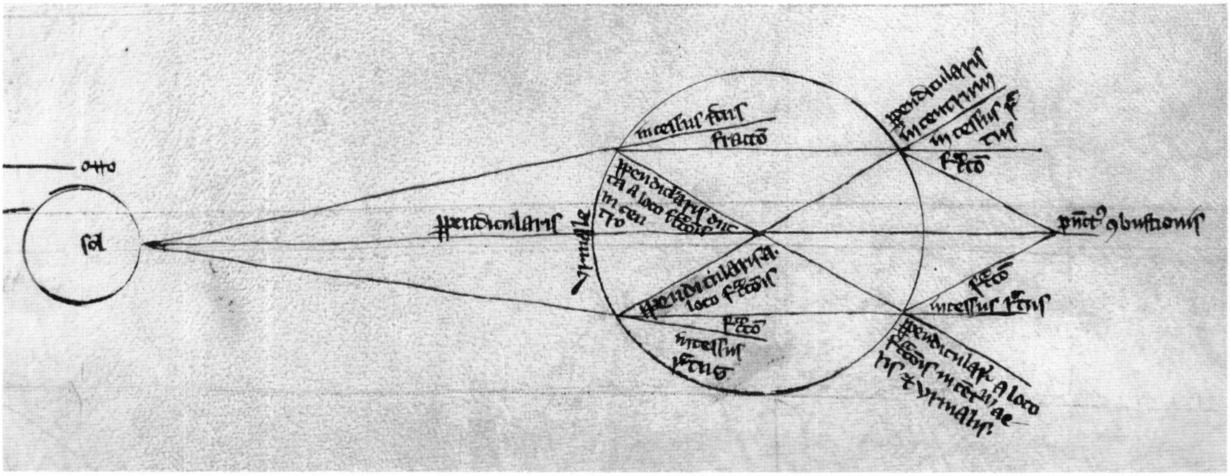

The key to experimentation, its advocates argued, was that one could generalize from natural processes to those observable under controlled conditions in the laboratory. Such reasoning, for example, led William Gilbert at the start of the seventeenth century to propose that the Earth itself is a kind of magnet. Historian of science David Wootton says that in Gilbert's book *De magnete* (1600), "For the first time the experimental method had been presented as one capable of taking over from traditional philosophical inquiry and transforming philosophy." It was in the seventeenth century that scientists (although they were not called that for another two hundred years) began to work in ways that the modern scientist would recognize: the time traditionally called the "scientific revolution," although that's a highly contested term today.

Ernest Board's twentieth-century oil on canvas of William Gilbert demonstrating a magnet to Elizabeth I in 1598, Wellcome Collection, London. Gilbert's *De magnete* was regarded as the leading work on magnetic and electrical phenomena at the time.

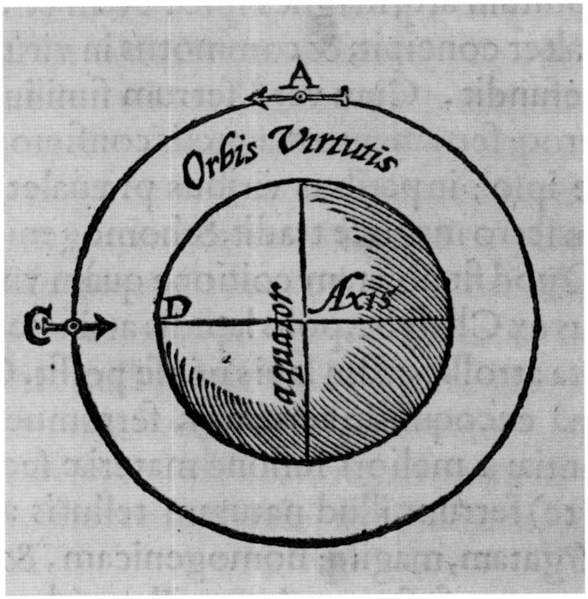

Illustration from William Gilbert's *De magnete*, London: Printed by P. Short, 1600, Wellcome Collection, London. Gilbert believed the Earth to be a kind of magnet.

A shift of such magnitude doesn't just happen by everyone changing the way they work. It also demanded a change in the *culture* of science: an acceptance that this is the right way to study nature. As Wootton explains, "what marks out modern science is not the conduct of experiments, but the formation of a critical community capable of assessing discoveries and replicating results." This meant, for example, agreeing on the right way to present your findings (namely, in the kind of objective, impersonal accounts scientists still produce today). It also meant agreeing on what and whom to trust. The Royal Society, founded in 1660 in London and inspired by Bacon's vision of empirical science, built up an international network of trusted sources of experimental reports. Reputation (and often social status) mattered; science became a socially negotiated affair, and you had to play by the rules.

This aspect of science comes out loud and clear in a survey of its history of experimentation. Who has the resources to conduct experiments? Whose voices are heeded (and whose are ignored)? How is an experiment made persuasive? Who are its target audiences? The rise of experimental science brought reliable knowledge verified by experience, but not without complications and caveats created by its social context. The history of science (experimental or not) is no paradigm of diversity, although historians today are rescuing contributions previously neglected, forgotten, or ignored. We can hope that the science of the future will do better, and will be the better for it.

Experiments testify to science's embrace of ignorance. Arguably the worst thing a scientist can do is to suppose they know what will happen in a given scenario without bothering to check. The rise of the experimental philosophy coincided with the liberation of curiosity as a valuable rather than a questionable attribute. For all that experimental science today is often assumed to be undergirded by a philosophical framework and an approved methodology ("state your hypothesis and then test it"), the fact is that, as philosopher of science Ian Hacking says, "One can conduct an experiment simply out of curiosity to see what will happen." Indeed, in the view of Charles Darwin's son, the astronomer George Darwin, once in a while one should do a completely crazy experiment, even if it is most likely to prove fruitless. You never know until you try.

James Gillray's 1796 etching with watercolor, *A Natural Philosopher,* depicting Adam Walker performing scientific experiments, Wellcome Collection, London.

X

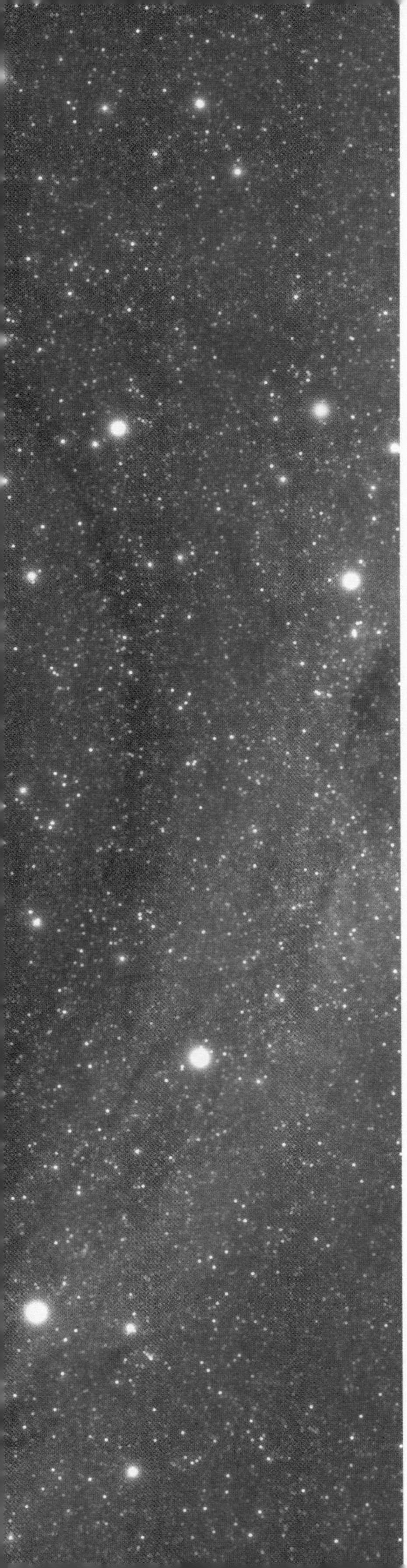

CHAPTER ONE

How does the world work?

01

Measuring the size of the Earth (third century BC)

Q What is the Earth's radius?

Contrary to common belief, most scholars since antiquity have recognized that the Earth is round—or, more properly, that it is (almost) spherical. That, said Aristotle in around 350 BC, explains why travelers see different stars at different locations, and why the shadow cast on the Moon by the Earth during an eclipse is curved. Aristotle adds that those who have attempted to measure the size of the Earth (he does not specify how) put its circumference at around 40,000 miles—impressively close to today's accepted value of 24,900 miles for the equatorial circumference.

The earliest determination of the size of the Earth for which we have an actual record was made by a Greek mathematician named Eratosthenes, who made his estimate about a century after Aristotle. He was a contemporary and friend of Archimedes, who asserted (again without explanation) that the Earth is about 30,000 miles around. Eratosthenes's estimate was even more accurate—by some calculations (for the conversion from Ancient Greek units is not straightforward) within just a few percent of the real value—and was accepted for many centuries, for anyone could see the elegant reasoning by which he got his result.

No copies remain of Eratosthenes's book *Measurement of the World*, so we must rely on what others recorded about his work. We don't know exactly how Eratosthenes deduced his method, nor what measuring device he used to put it into practice. But the principles are clear enough. They begin with the observation, well known in the ancient world, that—at all locations along the line of latitude called the Tropic of Cancer—the Sun stands exactly overhead on the summer solstice, so that shadows disappear (more strictly, they fall vertically). One such place this occurs is the Egyptian city of Aswan, which was then called Syene. The same applies (six months apart) to the Tropic of Capricorn, south of the equator, and it is a consequence of the tilt of the Earth's axis of rotation relative to its orbit, but Eratosthenes did not know that, or need to.

Everywhere north of the Tropic, shadows on the solstice fall at an angle. Eratosthenes realized that, from the ratio of the length of a shadow at a place due north of Syene—say, Alexandria—to the distance between the two locations, he could deduce the Earth's circumference. At least, this is so if one assumes that the Sun's rays arrive at the Earth parallel to one another—which is more or less true provided that the Earth is very small relative to the whole cosmos, as Aristotle claimed.

To put it another way: the angle at which the shadow of a vertical rod falls in Alexandria on the solstice is the same as the angle between lines running from the Earth's center to Alexandria and to Syene. By measuring this angle, Eratosthenes could work out what fraction of the Earth's circumference corresponds to the distance between the two cities. The argument relies on simple geometry of the kind described in Euclid's famed treatise, *Elements*.

It was a simple measurement in principle, most probably done using a Greek sundial in which the gnomon (rod) stands in a bronze bowl. In practice, the measurement involved several approximations—not least, that Syene is not exactly on the Tropic, nor is Alexandria exactly due north of it. Still, Erastothenes's result is astonishingly good.

Is this even an experiment—or simply an observation? Eratosthenes didn't use any ingenious apparatus—he could just as well have used a pole or tree—nor did he perform any manipulation, but

ERATOSTHENES
CA. 276–CA. 195 BC

Born in Cyrene in modern-day Libya, Eratosthenes was educated in Athens. He seems to have had all-encompassing interests, including mathematics, poetry, and history. He wrote a book (also lost) on geography, which advocated the use of a grid of what we'd now call lines of latitude and longitude. At the invitation of the Egyptian pharaoh Ptolemy III he became chief librarian of the famed Alexandrian library, and tutor to his children.

See also: Experiment 2, Direct demonstration of the rotation of the Earth, 1851 (page 16).

only made measurements of what he saw. Yet his work is not merely descriptive, like reporting a new species of animal, for example. He collected data and used them to deduce something quantitative and not obvious to the eye. That's often the way with experiments: they enable deductions beyond what is actually measured. In the end, where the boundaries lie between "experiment" and "observation" are a matter of taste.

What's more important here is the reasoning involved. Eratosthenes assumed he could use the familiar principles of geometry—literally "measuring the earth," for example in surveying and architecture—to measure the heavens. That might seem trivial today, but for the ancient Greeks it was by no means obvious that the celestial realm was governed by the same principles as the terrestrial. As philosopher of science Robert Crease puts it, "Eratosthenes came up with the audacious notion that the same techniques that had been developed for building houses and bridges, laying out fields and roads, and predicting floods and monsoons, could provide information about the dimensions of the earth and other heavenly bodies." Given that notion, one could work out ways to estimate other cosmic dimensions, such as the distance to the Moon or the stars. We could begin to locate ourselves in the universe.

It was a leap of reasoning of the same sort that led Isaac Newton to connect the trajectories of objects (like apples) falling on Earth to the orbits of the planets. Such an assumption of universality—of laws shared throughout the cosmos—lies at the heart of science.

The Well of Eratosthenes at Syene. From *The Adolfo Stahl lectures in astronomy ... 1916–17 and 1917–18*, Astronomical Society of the Pacific, San Francisco: Printed for the Society by the Stanford University Press, 1919, Plate XXXI, Fig. 1, University of California Libraries.

Direct demonstration of the rotation of the Earth (1851)

Q Can the rotation of the Earth be demonstrated from how objects move?

The Earth's rotation makes the Sun appear to rise, cross the sky, and set. But what is really moving? Until the heliocentric theory of the Polish astronomer Copernicus in the sixteenth century, which set the Sun at the center of the cosmos, it was generally thought that the Earth is stationary and the Sun orbits around it.

By the nineteenth century, Copernican theory was accepted by all scientists: the Earth orbits the Sun and rotates on its axis to bring about the diurnal cycle of day and night. Yet no direct demonstration of that axial rotation could be inferred by, say, a person in a windowless room.

That changed in 1851 with the famous pendulum experiment of Jean-Bernard-Léon Foucault in Paris. Simply by watching the swinging of a huge pendulum, Foucault showed, we can deduce that the Earth is turning: as he put it, the experiment speaks "directly to the eyes."

In the late 1840s, Foucault became interested in the technique of photography invented by Louis Daguerre in Paris. Working with his colleague Armand-Hippolyte-Louis Fizeau, he made photographic images of the night sky. But as the stars appear to move across the sky due to the Earth's rotation, the long exposure times needed by the plates meant the stars became smeared into streaks. To compensate, Foucault created a pendulum-driven clockwork instrument to realign the camera smoothly as the pendulum oscillated. While experimenting with this device in 1850, he saw that it seemed to rotate very slowly of its own accord. Repeating the experiment with a simple pendulum—a weight suspended with piano wire—he saw the same thing. He realized the apparent movement of the pendulum's plane of swing was not, in fact, that at all; what was moving was everything else around it as the Earth slowly turned.

This should have been no surprise to anyone. It was long known that the rotation of the Earth should have observable consequences for the trajectories of objects not stuck, as it were, to the planet's surface. A weight dropped from a great height will not fall to the spot directly below it because the Earth moves slightly during the descent. Similarly, the Earth's rotation should

Foucault's pendulum in the Panthéon, Paris, by which he demonstrated the rotation of the Earth.

LÉON FOUCAULT | 1819–1868

Jean-Bernard-Léon Foucault had no formal training in physics. He began studying medicine before deciding he was too squeamish to be a surgeon, and then he became a journalist. He had an enduring fascination with mechanics and inventions, and his work on celestial photography led him to research mirrors and telescopes. After his celebrated demonstration of his pendulum in the Panthéon in Paris, Napoléon III appointed him a physicist at the Paris Observatory.

See also: Experiment 1, Measuring the size of the Earth, third century BC (page 14); Experiment 34, The wave nature of light, 1802 (page 150).

slightly deflect the impact point of a cannonball fired into the air. Many experiments on falling and flung objects had sought to observe this effect, but it is so tiny that it had never been seen.

The effect of the Earth's rotation on pendulums was also recognized. The French scientist Siméon Denis Poisson had described it in 1837 but deemed the effect too small to be detectable. In fact, Galileo's pupil Vincenzo Viviani seems to have observed the effect two hundred years earlier, but he considered it merely a nuisance and didn't connect it with planetary rotation. To see the rotation indeed requires great care. It can be disturbed by air resistance on the swinging bob and also by stray air currents and friction at the point of attachment at the top of the wire. The longer the wire and the heavier the bob, the less important these confounding factors. In January 1851, Foucault first saw the expected rotation with a 6½-foot-long pendulum suspended in the basement of his Paris home. "Everyone who is in its presence," he wrote, "grows thoughtful and silent for a few seconds, and generally takes away a more pressing and intense feeling of our ceaseless mobility in space."

At the invitation of the director of the Paris Observatory, he repeated the demonstration in the observatory's central hall with a wire 36 feet long. After he reported his results to the French Academy of Sciences in February, Emperor Napoléon III asked him to stage a public demonstration, which he did in the great church of the Panthéon, using a 61-pound bob attached to a 220-foot wire less than 1.5 millimeters thick. Around the circumference of the swing were stationed two semicircular banks of wet sand; a pointer attached to the bob cut out marks in the sand, the positions of which moved with each stately swing. It was a sensation, and the experiment was soon replicated in institutions around the world.

Foucault used trigonometry to show that the amount of (apparent) rotation of the pendulum's plane varies with geographical latitude. Only at the poles does it execute a full 360-degree revolution, as demonstrated in 2001 by scientists in Antarctica.

Foucault's pendulum experiment. Engraving from William Henry Smyth's *The Cycle of Celestial Objects continued at the Hartwell Observatory to 1859*, London: Printed for private circulation by J. B. Nicols and Sons, 1860. Held in a private collection.

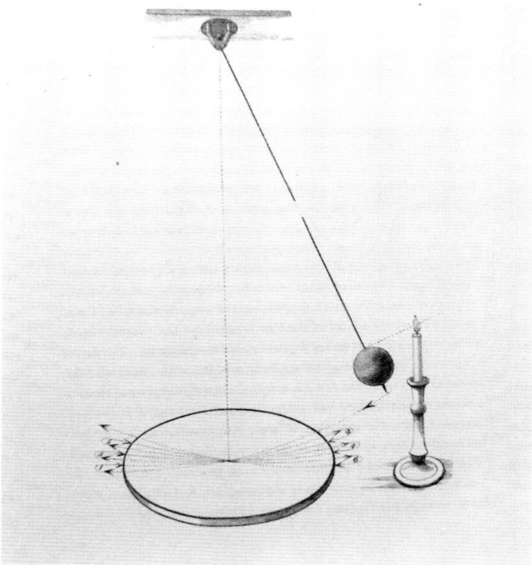

Attempting to detect the ether (1887)

Q Can the relative motion of the Earth in the ether be detected?

Since the time of Isaac Newton, the universe was imagined as being populated by objects located and moving on a fixed backdrop of space, every position having precise coordinates. Light was thought to move through this space as waves of oscillating electric and magnetic fields in an all-pervasive fluid called the *luminiferous* (light-bearing) ether, which was too tenuous to be detected directly. In 1880 Albert Michelson, an instructor in physics at the Naval Academy in Annapolis, Maryland, figured that he could demonstrate the ether's presence by measuring the speed of light traveling in two directions at right angles to one another. Two years earlier, Michelson had made the most accurate measurement to that date of the speed of light, deducing it to be 186,380 miles per second.

As the Earth progresses along its orbit around the Sun, Michelson figured, it moves through the ether, creating an "ether wind." A light beam sent out in the direction of the Earth's motion should be slowed by this wind, while another beam at right angles to the motion wouldn't be affected. Michelson compared this to the way a person is slowed down by swimming upstream in a river relative to swimming the same distance crossways from bank to bank.

To detect the tiny change in light speed as it moves "into the wind," Michelson decided to seek interference effects between two perpendicular beams as they bounce back from mirrors and intersect. If they start out with the peaks of their waves in step—ensured by splitting a single beam into two perpendicular ones using a mirror that was partly transparent and partly reflective—then traveling the same distance but at slightly different speeds would bring the peaks slightly out of step, causing a partial canceling out of one light beam by the other and producing a series of light and dark bands. This kind of measurement is said to be interferometric—"measuring interference."

In 1882 Michelson joined the Case School of Applied Science in Cleveland, Ohio. On a train trip he met the chemist Edward Morley of Western Reserve University in nearby Hudson, and the two agreed to collaborate on the experiment. But before they could carry it out, Michelson seems to have had a mental breakdown—his marriage disintegrated and he received psychiatric treatment in New York. It wasn't until 1887 that Michelson and Morley were ready to conduct the test.

ALBERT MICHELSON 1852–1931

Albert Abraham Michelson was the first American scientist to win a Nobel prize. Born in Poland (then part of Prussia), he was awarded a place at the US Naval Academy in Annapolis, Maryland in 1869, by order of the American president Ulysses S. Grant himself. After working at the Case School of Applied Science in Cleveland, Ohio, he became the founding head of the physics department at the University of Chicago in 1892.

See also: Experiment 34, The wave nature of light, 1802 (page 150).

It demanded incredible precision—even vibrations from carriages passing outside the lab could disturb the apparatus and cause errors. The two researchers built their interferometer —a lamp and a series of mirrors that produce and reflect the perpendicular beams—on a slab of sandstone 5 feet square, floating on a wooden support in a trough of liquid mercury to damp out vibrations. Michelson observed the interference bands through an eyepiece without touching the apparatus, calling out the results to Morley. The experiment set new standards of precision, but the results seemed disappointing: they could detect no difference in the speed of the two light beams. The ether remained invisible. The researchers concluded that, instead of simply moving through the passive ether, the Earth must be dragging some of that fluid along with it, so there was no detectable wind after all.

There was another way to explain the apparent failure, though. Dutch physicist Hendrik Lorentz argued that electromagnetic fields, like those of a light wave, are deformed when they move, producing a contraction of the wavelength of light that exactly compensates for the change in light speed through the ether due to the Earth's motion.

In 1905, Albert Einstein turned that argument on its head. What if, he asked, the "Lorentz contraction" isn't changing the light waves, but is, in fact, a deformation of space and time itself? In other words, the speed of light stays the same but relative motion alters the very backdrop of space. If that's so, Einstein showed, there is no longer any need to invoke a luminiferous ether at all. This is the basis of Einstein's theory of special relativity, which showed that space and time are not absolute but are altered for two objects by their relative motion. Although it's not clear how much Einstein's thinking was influenced by the Michelson–Morley experiment, Michelson was awarded the 1907 Nobel Prize in Physics for his high-precision measurements on light. As he showed, even an experiment that produces a "null result"—no sign of what you expect—can transform our understanding of the world.

EDWARD MORLEY | 1838–1923

Edward Williams Morley spent his academic career as a professor of chemistry at Case Western Reserve University in Cleveland, Ohio. Despite his chemical training, he also made significant contributions in physics and optics.

See also: Experiment 35, Measuring the speed of light, 1849 (page 152).

The Michelson–Morley interferometric setup mounted on a stone slab floating in an annular trough of mercury, 1887, Albert A. Michelson papers, Manuscript 347, Nimitz Library, US Naval Academy, Annapolis, Maryland.

Testing general relativity (1919/1959)

Q Is Einstein's theory of general relativity correct?

Einstein's theory of relativity, which shattered traditional "common-sense" notions of space and time, came in two instalments. Special relativity, which he described in 1905 and is the source of Einstein's iconic equation $E=mc^2$, considers what happens when objects move very fast, concluding that space is shortened and time is stretched. General relativity, unveiled in 1916, considers objects that *change* their speed: in other words, that accelerate. Bodies accelerate as they fall to earth, and also as they move in circles, like the Moon orbiting the Earth. Both motions are caused by the force of gravity, and Einstein revealed a deep connection between gravity and acceleration. He figured that what we call the gravitational force is actually the result of distortions in the very fabric of space and time that cause objects to accelerate or deviate from moving in a straight line.

Those distortions are caused by mass. Think of a heavy person standing in the center of a trampoline. As a ball rolls across the surface, its path is bent by the depression, and maybe it spirals inward to the person's feet. As physicist John Wheeler later put it, mass tells spacetime how to curve and the curved spacetime tells massive objects how to move.

This was a very different picture to the idea of gravity that had prevailed previously, developed by Isaac Newton in the seventeenth century. In Newtonian theory, gravity was a force, like magnetism, that acted through space. But for Einstein, gravity was an apparent force caused by space as it was reshaped by massive objects.

Many other scientists found this an outlandish idea. However, general relativity could be tested, because it made predictions. For example, the distortions of space close to the immense mass of the Sun would make the orbit of the planet Mercury move slightly with each revolution in

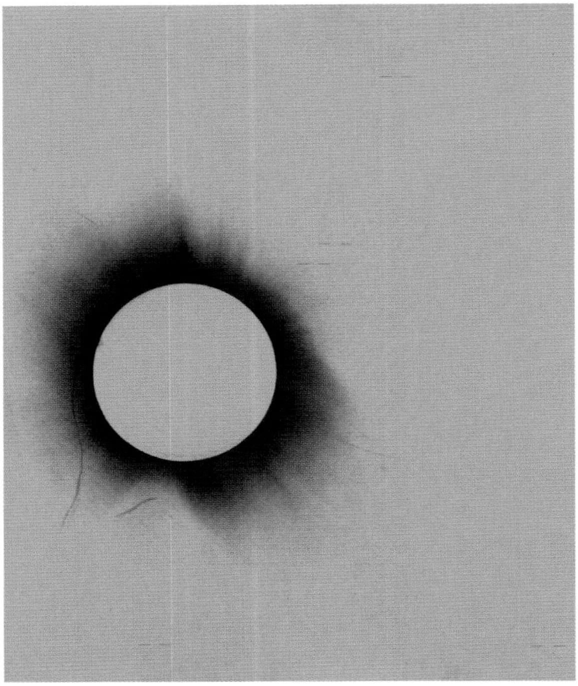

Solar eclipse reproduced by Arthur Eddington and Frank Dyson. From "A determination of the deflection of light by the sun's gravitational field, from observations made at the total eclipse of May 29, 1919," *Philosophical Transactions,* Series A, London: Royal Society of London, January 1, 1920, Vol. 220, Plate 1, Natural History Museum, London.

a way that Newtonian gravity cannot explain. Astronomers had already seen such an anomalous shift of Mercury's orbit and were puzzled by it. But it exactly matched what general relativity forecast.

Still, a theory this bizarre needed more proof than that. One of the remarkable consequences of Einstein's theory was that light might not travel in straight lines through empty space, as had always been thought. If the light rays pass close to a very massive object, the distortions of space will bend their path. This effect should be evident in starlight passing close to the Sun on its journey to Earth: those stars would appear shifted in the sky compared to when viewed at night. The apparent displacement of the stars would be tiny, but perhaps big enough to see using a large telescope.

However, you can't see stars when the Sun is in the sky because they are much too dim. Unless, that is, the Sun's light is blocked by the Moon during a total solar eclipse.

The British astronomer Arthur Eddington, director of the observatory at Cambridge University, was eager to conduct this experiment. Eddington was one of the most enthusiastic advocates of Einstein's theory in England, where many of his colleagues were sceptical.

A total solar eclipse on May 29, 1919 supplied a perfect opportunity to test Einstein's theory. It would be observable close to the equator, and plans for an astronomical expedition to make the observations were drawn up in 1917 in the midst of the First World War. Eddington's team decided that the best observing locations were in northern Brazil at a place called Sobral and on the island of Principe, off the coast of what is today Equatorial Guinea. Eddington would lead the African party, while others would go to Brazil.

Eddington's team arrived at Principe to find an island plagued by rain, mosquitoes, and mischievous monkeys. The weather got worse as May progressed, and the day of the eclipse began with a heavy downpour. But in the nick of time, just as the sky went dark, the clouds parted enough for Eddington and his team to get a few halfway decent photos of the stars, which blinked into view around the darkened Sun.

ARTHUR EDDINGTON 1882–1944

As an astronomer at Cambridge University, Arthur Eddington made his name as an expert on the structure of stars. He was also a deep thinker on the philosophy of science, as well as a gifted communicator whose 1928 book *The Nature of the Physical World* sought to introduce the "new physics" of relativity and quantum theory to a general audience. His eagerness to test and hopefully verify Einstein's theory was not purely scientific. As a Quaker, he was committed to peace and reconciliation between nations after the war, and he hoped that his eclipse experiment would foster good Anglo–German relations and show how science can transcend political divisions and conflicts.

See also: Experiment 6, The discovery of gravitational waves, 2015 (page 28); Experiment 35, Measuring the speed of light, 1849 (page 152).

The team in Brazil, on the other hand, enjoyed wonderfully clear skies for their observations. It was only afterward that they discovered the Sun's heat had warped the main focusing mirror of the telescope and the most important photographic plates were blurred and all but useless.

Could enough data be salvaged from the observations to test Einstein's theory? When Eddington arrived back in England, he and the Astronomer Royal Frank Dyson pored over the only two photographic plates from Principe that showed enough stars to reliably detect any shift of those close to the edge of the Sun. To their relief, they found that some of the images made with a small backup lens in Brazil were useable too. In November, Eddington and Dyson announced to a crowded meeting of experts and journalists that the stellar positions had indeed shifted. Einstein's theory was vindicated.

Composite illustration of two photographs taken by Arthur Eddington's expedition at Sobral, in northern Brazil, showing the deflection of starlight by the Sun, *Splendour of the Heavens*, London, Hutchinson & Co., 1923, University of Illinois Urbana-Champaign.

Eddington's discovery made headlines around the world: in London *The Times* hailed it as a "Revolution in science." As Eddington wrote to Einstein at the end of the year: "All England is talking about your theory."

General relativity was a new theory of the entire universe. When Einstein used this theory to calculate the shape of that universe, he found it made an odd prediction: the universe was expanding. No one at the time thought this could be true, and Einstein added a "fudge factor" to his equations to get rid of the expansion. But only a few years later, astronomers discovered that the universe is indeed getting bigger—propelled, as we now understand, by the Big Bang in which it began. (Began from what? No one is yet sure.)

After Eddington's eclipse observations, resistance to Einstein's theory steadily faded. But an idea this profound deserves plenty of empirical testing, and scientists have been finding new ways to do that ever since. One especially elegant test was proposed in 1959 by Canadian–American physicist Robert Pound of Harvard University and his graduate student Glen Rebka. Since general relativity predicts that time runs more slowly in stronger gravitational fields, a clock in space should be a little faster than one on the Earth's surface. One of the most accurate of "natural clocks" is provided by the vibration frequencies of electromagnetic radiation such as light.

Atoms emit and absorb light at well-defined frequencies. The light emitted by one atom should therefore have just the right frequency to be absorbed by another identical atom. But if the atoms are in gravitational fields of different strength, gravitational time distortion means that the frequency emitted by the first atom will no longer match that of the absorbing atom by the time it reaches it. The effect would generally be tiny, but Pound and Rebka calculated that it should be detectable if the emitting atom—they used a radioactive form of iron that emits gamma rays—is at ground level and the absorbing atom is located at the top of a high tower.

Pound and Rebka performed the experiment in 1959 using a 74-foot tower at the Jefferson Physical Laboratory at Harvard. Around the vertical path of the gamma rays the researchers fitted a cylindrical polymer bag filled with helium, so that the rays were not absorbed by air. They could compensate for the gravitational frequency shift by moving the absorber up and down using a vibrating loudspeaker cone, producing another frequency shift via the Doppler effect. In this way, the absorber and emitter moved in and out of resonance, changing the amount of absorption detected. The change was within 10 percent of the value predicted by general relativity—an accuracy that the researchers narrowed to 1 percent by improvements to the experiment in 1964.

Glen Rebka at the bottom of a 74-foot column filled with helium, which was used to test Einstein's general theory of relativity, at the Jefferson Physical Laboratory, Harvard, in 1959.

Experiments to test Einstein's theory of general relativity are still taking place, both in space and in the laboratory. It's not that anyone wants to prove Einstein wrong, but physicists are convinced that, just as general relativity is the theory of which Newtonian gravity was just an approximation, so there should be a theory beyond general relativity that reconciles it with other aspects of physics. Any deviations could be the vital clues to what that new theory might look like.

What is an experiment?

The notion of an "experiment" meant different things at different times. Even today, the nature of the interplay between experiments and theories remains contentious. Some scientists feel that theories that cannot be experimentally tested (and potentially falsified) don't qualify as genuine science. This idea that science consists of exposing hypotheses to potential falsification by experiment was articulated most clearly by the philosopher Karl Popper in the mid-twentieth century. It plays an important role in experimental science, but that role is often exaggerated.

For one thing, hypotheses have not always been welcomed—in early-nineteenth-century England they were regarded as suspiciously speculative. Besides, proving hypotheses wrong is complicated (a point of which Popper was well aware). If an experiment produces a different result to the one you predict, does that mean your theory is wrong? Maybe it just means that part of an instrument is broken or isn't performing as intended; maybe you overlooked another factor influencing the outcome; or maybe the samples were contaminated. Or maybe the theory needn't be jettisoned, but simply modified. Being human, scientists are far more inclined to tweak their pet theory in the hope of finding a closer match with unanticipated results than to ditch it outright—perhaps a good thing, as science might be too brittle if every contradictory experiment was regarded as a falsification.

Popperian falsification does, however, dispel the idea that scientists observe some phenomenon and then cook up a theory to explain it. You must have a theory *first* to formulate a prediction at all. It is the theorist, said Popper, "who shows the experimenter the way."

As the French physicist and philosopher of science Pierre Duhem pointed out, an experiment is virtually an embodiment of theory, in that the things you measure (and those you ignore) are themselves based on theoretical concepts: physicists see the displacement of a needle on a dial and decide this measures a quantity like "force" or "charge," say. In consequence, said Duhem, if a prediction is falsified by experiment, we can't know if the fault lies with the theory being tested or with all the other beliefs and hypotheses it rests on: the problem is "under-determined." This predicament makes science a complex web of interrelated hypotheses and greatly complicates its empirical basis. Philosophers still argue about how serious an issue this is. And what do you do if an experimental result is predicted equally well by two theories, with no obvious way to choose between them?

So, even if something evidently happens (or fails to happen) in an experiment, arguably it only becomes a "result" when we know what to do with it. For example, it was very hard to know "what to do" with the failure of Michelson and Morley to observe the light-bearing ether (see page 18) until Einstein showed (and others agreed) that no ether was needed after all. In this respect, says philosopher of science Ian Hacking, "One could say that the experiment lasted half a century."

Does all this mean that experiments themselves are never neutral and unbiased, and their interpretation never unique or incontrovertible? To some degree, yes, but that is not a problem if we accept that experimental science is, in fact, a delicate dance between theory and observation, in which neither is dominant. What perhaps matters more is that because, as Duhem points out, there is a near-infinity of conceivable experiments that can test a given hypothesis, the art is to identify *good* experiments: those that have the greatest power to test the idea. It also means that two experiments that might look utterly different to non-experts might *mean* the same thing to experts.

So what makes for a good experiment? Science has always been plagued by false or unverified experimental claims. For example, in 1988 scientists led by French immunologist Jacques Benveniste reported that chemical solutions of a biological agent continued to show biological activity even when diluted well beyond the point where a single active molecule remained. Benveniste believed this showed water has a "memory"—retaining an imprint of molecules dissolved within it. Although reported in good faith, the results could never be replicated and are now regarded as an example of what American chemist Irving Langmuir called "pathological science."

It is arguably better to regard the results as poor experiments. The way they were designed precluded the likelihood of an unambiguous outcome. There were too many uncontrolled factors that might influence the results. The art of scientific experimentation consists largely in making it discerning: finding a scheme through which your hypothesis can be probed stringently and potentially ruled out decisively.

Scientists often assert that their practice is governed by the "scientific method," in which one formulates a hypothesis that makes predictions and then devises an experiment to put them to the test. But this is a modern view, codified in particular by the "pragmatist" philosophers of the early twentieth century like John Dewey and Charles Sanders Peirce. Later philosophers of science such as Paul Feyerabend question whether science has ever been so formulaic and argue that its ideas depend as much on rhetorical skill and persuasion as on logic and demonstration.

That unsettles some scientists, who insist on "experience"—observation and experiment—as the ultimate arbiter of truth. But although in the long run a theory that repeatedly conflicts with experimental observation can't survive, in the short term theorists may be right to stick to their guns in the face of apparent contradiction. More often, supporters of rival theories might argue about the interpretation of an experiment. One party may triumph not because their interpretation is right but because they're better at presenting their case. Or a scientist may reach the wrong conclusions from a correct and even elegant experiment just because they posed the wrong question. All this makes the scientific enterprise complicated, ambiguous, and socially contingent, but also richer, more creative, and more gloriously human.

Experimental methods: here, French physiologist Jean-Paul Langlois sets up an experiment to study a cyclist in 1921.

The violation of parity (1956)

Q Does nature distinguish between left and right?

The reason why we so often confuse our left and right is because they are equivalent: each is the mirror image of the other. This is an example of what scientists mean by a fundamental symmetry of nature. Left and right are related by symmetry in a way that up and down are not, because gravity breaks that symmetry: objects only fall down, not up.

Symmetry relationships lie at the core of the most fundamental theories of physics. The laws of physics are the same in New York as in Shanghai, because they have translational symmetry: a shift ("translation") in space makes no difference. The laws of mechanics are also time-symmetric: they would work the same way if time could be run backward. And the laws of physics were long considered to be invariant to left-right switching too: they will work the same way in a looking-glass world.

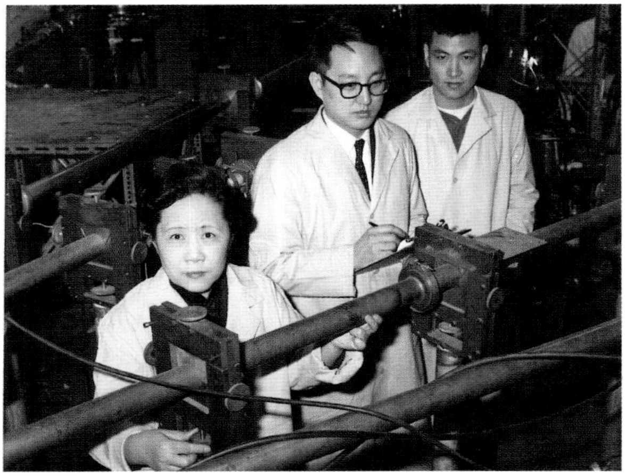

Chien-Shiung Wu (left) pictured with Columbia University physicists Y. K. Lee and L. W. Mo in 1963, Smithsonian Institution Archives, Washington, DC.

However, in the spring of 1956 two Chinese–American physicists, Tsung-Dao Lee of Columbia University in New York and Chen Ning Yang of the Institute for Advanced Study in Princeton, suggested that not all of physics might be indifferent to left-right inversion. That indifference is known in physics as *parity*, and Lee and Yang identified a physical process that they thought might violate the principle of parity. If that were so, then it would be the first indication that left and right were not wholly equivalent in nature.

The two physicists were led to suspect as much as they attempted to understand the properties of certain elementary particles involved in the so-called weak interaction. This is a force that acts inside atomic nuclei and gives rise to the type of radioactivity called beta decay, where a neutron spontaneously splits into a positively charged proton (which remains in the nucleus) and a beta particle (an electron, which is ejected).

Many physicists were sceptical of the idea, but Yang and Lee decided to it test it experimentally. Lee knew that his colleague at Columbia, the Chinese–American scientist Chien-Shiung Wu, was an expert in beta decay, and so they approached her with the suggestion. Wu believed it was worth a shot—and figured she could see a way to do it.

The concept was simple. Atomic nuclei have a quantum-mechanical property called spin, which doesn't exactly mean the same as the everyday term but is similar. Quantum spin makes the nucleus act like a little magnet, with poles that point in opposite directions. Wu knew that the

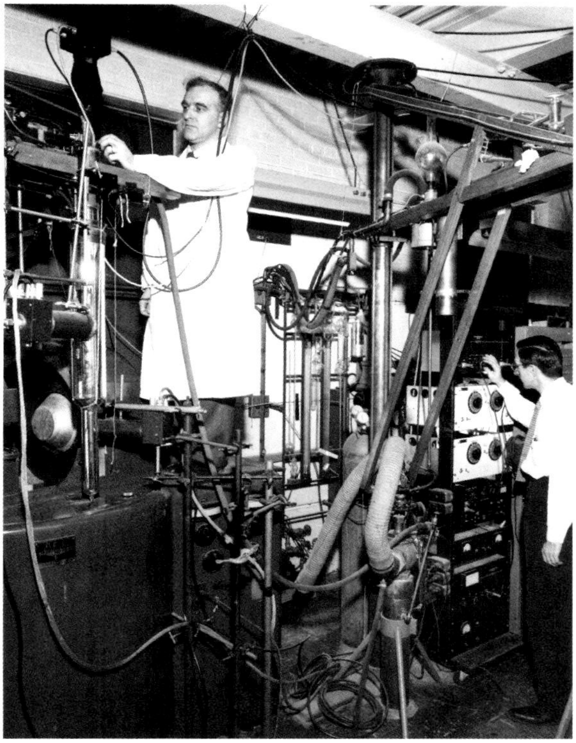

Ernest Ambler and Raymond W. Hayward with the parity experiment apparatus at the National Bureau of Standards in 1956, National Institute of Standards and Technology Digital Archives, Gaithersburg, Maryland.

CHIEN-SHIUNG WU | 1912–1997

Born in Shanghai, Chien-Shiung Wu gained a place to study nuclear physics at the University of California at Berkeley, at that time regarded as the mecca of the discipline. There she worked in the late 1930s with Ernest Lawrence, the inventor of the cyclotron particle accelerator. After completing her PhD in 1940, she became the first female instructor in the physics department of Princeton University. She moved to Columbia University in New York, and in 1944 was brought into the Manhattan Project as an expert on nuclear decay. Wu's famous experiment established her reputation for a determined yet rigorous approach to experimentation, which she went on to demonstrate in further studies of beta decay and other aspects of nuclear science. She was a staunch advocate for women in science and spoke out against the obstacles created by bias and discrimination.

See also: Experiment 18, The three-dimensional shape of sugar molecules, 1891 (page 84); Experiment 25, Measuring the charge on an electron, 1909–1913 (page 112).

spin orientation of a nucleus that emits a beta particle determines the direction in which the particle is ejected. So, if she could get a bunch of beta-decaying atoms to align their spins, left-right equivalence (parity) means that the same number of electrons should be emitted in both directions along the axis of alignment. But if parity is violated in beta decay, slightly more electrons should emerge in one direction than in the opposite.

It was neat in principle, but hard to arrange in practice. To line up many nuclear spins, Wu needed to cool down the respective atoms (she used the isotope cobalt-60) to within a whisker of absolute zero—specifically, to no more than 0.018°F (0.01°C) above that ultimate frigid state—so that the jiggling of heat didn't mess up the alignment. The work was done by Wu and her graduate student Marion Biavati in collaboration with a team at the National Bureau of Standards in Maryland.

The team had the experiment working by late 1956, and within a few weeks they confirmed Lee and Yang's prediction: nature could tell left from right after all. The result shattered one of the long-standing assumptions of fundamental physics—in Wu's view it heralded a "sudden liberation of our thinking on the very structure of the physical world." Her friend, the Austrian physicist Wolfgang Pauli, who had been sceptical of the prediction, put it more colorfully: it showed, he said, that God is left-handed.

The discovery of gravitational waves (2015)

Q Are gravitational waves real?

In Einstein's 1916 theory of general relativity, empty space itself becomes a kind of "fabric"—a four-dimensional spacetime—that is deformed by the presence of mass to produce the gravitational force. One of the predictions of the theory was that when very massive objects accelerate, they can create ripples in spacetime that radiate out like those from a pebble dropped in a pond. These ripples are called gravitational waves, and their passing momentarily stretches and compresses space.

To produce a gravitational wave (GW) of any significant strength would require a very extreme disturbance: something catastrophic happening to bodies of enormous mass. At first Einstein felt that GWs would simply be too faint ever to be detected, and in 1936 he and a colleague even argued (erroneously) that general relativity didn't predict such waves after all.

One potential astrophysical source of GWs was another of general relativity's predictions: black holes, created when very massive stars burn out and collapse under their own gravity. In theory, this collapse proceeds until all the star's mass is squeezed into an infinitesimally small point: a so-called spacetime singularity. The gravitational field around such a singularity is so strong that even light cannot escape, hence the "blackness."

For several decades, no one believed black holes could really form—they considered spacetime singularities to be just a mathematical quirk of the theory. But in the 1950s and '60s, general relativity enjoyed a renaissance, in part because researchers became more adept at handling the challenging mathematics, and black holes were taken more seriously. Their existence is now pretty much universally accepted, and astronomers have even obtained images of them, revealed by the ultra-hot gas and matter that surrounds and is sucked into them.

If two black holes come close enough to each other, they are predicted to orbit one another and spiral inward until they merge, releasing an unimaginable amount of energy, partly in the form of gravitational waves. A single such black-hole merger should briefly radiate more energy than that emitted by all the stars in the observable universe.

The long arms of the LIGO detector interferometer located at Livingston, Louisiana, May 19, 2015, courtesy of the Caltech/MIT/LIGO Lab.

RAINER WEISS | B. 1932

Born in Berlin, Rainer Weiss arrived in the USA as part of the Jewish exodus from Nazi Germany. He received his doctorate in physics from the Massachusetts Institute of Technology in 1962 and joined the faculty two years later. As a key participant not just of LIGO but of projects that measured the cosmic microwave background—the ubiquitous afterglow of the Big Bang—Weiss has been at the center of efforts to understand the fundamental nature of the universe.

Once the possibility of both black holes and GWs created by their merging was accepted, the search for these ripples in spacetime began. The first GW detectors were devised in the 1960s by American physicist Joseph Weber. He reasoned that the tiny distortion of space from a passing GW might be revealed by a piezoelectric material, which produces a voltage when squeezed. Weber looked for telltale signatures in aluminum bars coated with a piezoelectric substance suspended from a wire: a so-called Weber bar. It would be very hard to distinguish such a GW signal from some other source of vibration, but Weber reasoned that a GW passing over the Earth should generate an identical signal almost simultaneously in two detectors situated a long distance apart. Such a coincidence would be hard to explain any other way. In 1969 he claimed to have detected such a signal in two Weber bars separated by a distance of 621 miles, but no one could replicate the result.

At the same time, Weber and others considered another method of detection that used interferometry. This was basically a variant of the technique Michelson and Morley used to seek evidence of the ether in 1887 (page 18). The idea is to use light beams as extremely precise rulers that will detect a distortion in space. You split a light beam in two and send the beams along two channels at right angles. At the end of the channel they are reflected back by mirrors. When they reunite, the waves will still be in step and will interfere positively, reinforcing each other. But if a GW distorts spacetime more along one arm of the interferometer than the other, then the waves will be slightly out of step after their journey, and the interference will shift detectably.

That's the theory. But calculations implied that the sensitivity needed to detect a GW in this way was daunting. The detector would have to measure a change in the relative length of the paths of the two beams of less than the diameter

Gravitational wave signals detected by the twin LIGO observatories at Livingston, Louisiana, and Hanford, Washington, in September 2015, courtesy of the Caltech/MIT/LIGO Lab. The signals came from two merging black holes, each about thirty times the mass of our Sun and lying 1.3 billion light years away.

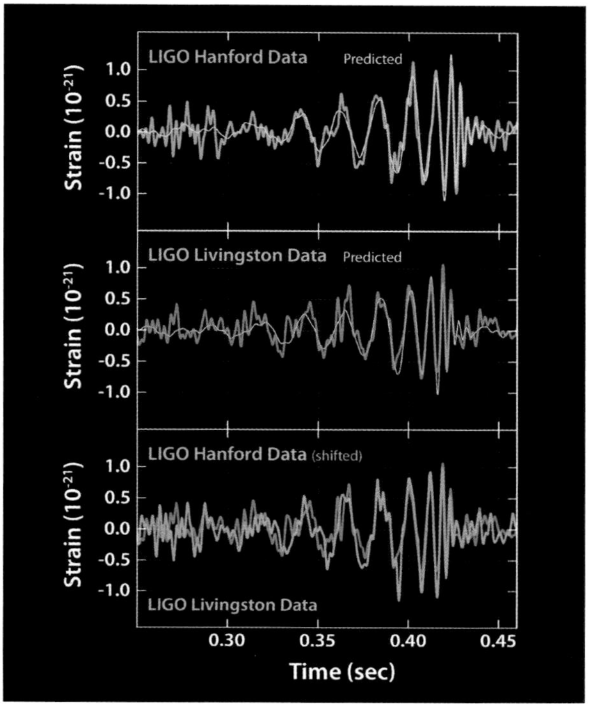

BARRY BARISH | B. 1936

A physicist at the California Institute of Technology, Barry Barish was chosen to head LIGO when construction began in 1994 because of his experience in running "Big Science" projects. He had headed projects in particle physics in the US and Italy, and so was no stranger to managing large teams dependent on huge and complex instruments. It is a skill increasingly valuable for pushing forward the frontiers of science.

of an atomic nucleus. The longer the arms, the greater the sensitivity.

In the face of such a challenge, many doubted that the method would ever work. But it was championed by Rainer Weiss, at the Massachusetts Institute of Technology (MIT), and theoretical physicist Kip Thorne of the California Institute of Technology (Caltech), who teamed up with Scottish experimental physicist Ronald Drever to start putting it into practice. At MIT Weiss made a prototype interferometric GW detector with arms a few feet long, and in the 1970s he extended the arms to a length of 29½ feet. The researchers knew that, to achieve the required sensitivity, ultimately the detectors needed to be huge, with arms several miles long. They joined forces to argue their case, and in 1984 funding was approved for a huge detector system called the Laser Interferometer Gravitational-wave Observatory (LIGO). As Weber had argued, the key to a convincing detection was to see

Computer simulation showing two merging black holes, the source of the gravitational waves detected on September 14, 2015 by LIGO, courtesy of the SXS (Simulating eXtreme Spacetimes) Project.

coincident signals in widely separated detectors. LIGO was to have two detectors based in the USA: one at Hanford, in Washington State, and one at Livingston in Louisiana, 1,864 miles apart. Each would have two arms 1¼ miles long; to produce a laser beam strong enough to sustain that journey demanded advances in laser technology.

Construction of LIGO started in 1994, led by physicist Barry Barish, and it began operating in earnest in 2002. Over the next two decades, the instrument was gradually upgraded and accumulated a team of around a thousand scientists from more than twenty countries. After a round of improvements completed in early September 2015, the souped-up LIGO had been operating for barely two weeks before both detectors saw a signal that looked exactly like the one predicted for a GW. The research teams deduced from the strength and shape of the signal that it had been produced by the merging of two black holes with masses of twenty-nine and thirty-six times that of the Sun, happening 1.3 billion light years away—which meant that the event took place while the only life on Earth was microbial. The announcement of the discovery in February 2016 caused a sensation, and led to a Nobel prize the following year for Barish, Thorne, and Weiss.

Although this was the first direct detection of a GW, indirect evidence had been discovered in 1974 when two US astronomers saw a system of two neutron stars gradually spiraling in toward each other as they radiated energy in the form of GWs. By the mid-2010s a second interferometric GW detector called VIRGO had been built near Pisa, Italy, and in August of 2017 the first joint detection of a GW was made by LIGO and VIRGO together.

There were several more GW detections in the years that followed, and now these spacetime ripples are opening a new window on the universe, revealing events that can't be studied in other ways. An immense space-based interferometric GW detector called the Laser Interferometer Space Antenna (LISA) is being planned by NASA and the European Space Agency, in which spacecraft will send, reflect, and measure the light beams along paths several million miles long.

LIGO and its successors exemplify the new era of experimentation in fundamental physics, in which huge and expensive instruments designed and run by teams of thousands of scientists and technicians probe nature at the very limits of what is detectable. What that implies for the nature of experimentation itself is mixed. There may be no other way of investigating such elusive features of the universe, but these projects are years in the planning and are also at risk of groupthink, having none of the nimbleness and improvisation of traditional benchtop science. They are not better or worse than the old ways of experimentation, but they are different.

KIP THORNE | B. 1940

Kip Thorne has been at the center of the reviving interest in general relativity since it began in the 1960s, when he was a student of the influential Princeton physicist John Wheeler. He is perhaps the closest thing the field has to a superstar after English physicist Stephen Hawking, having been an advisor on the 2014 movie *Interstellar* to make its depiction of a black hole scientifically accurate.

See also: Experiment 3, Attempting to detect the ether, 1887 (page 18); Experiment 4, Testing general relativity, 1919/1959 (page 20).

fig:3

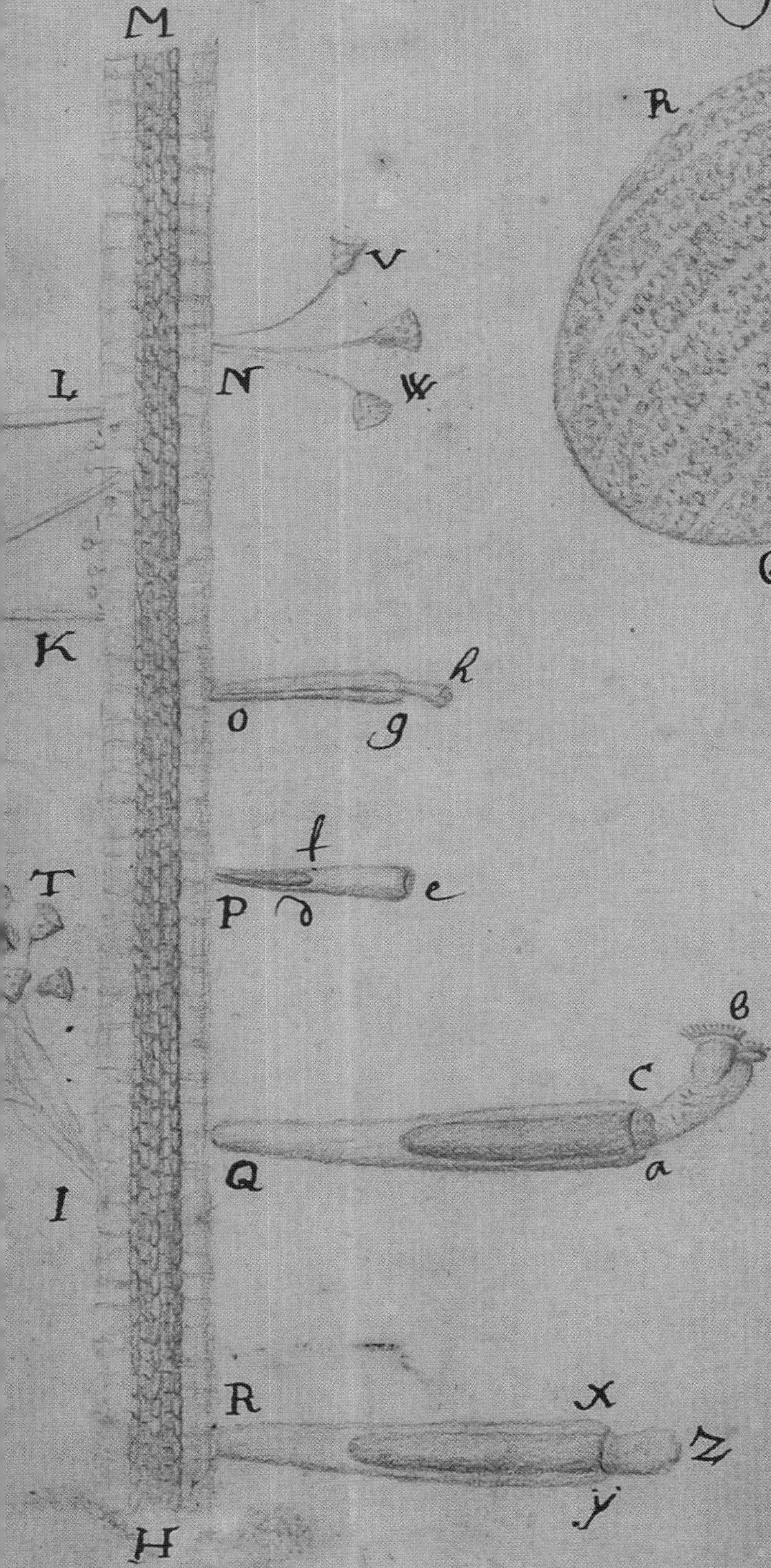

fig:7,

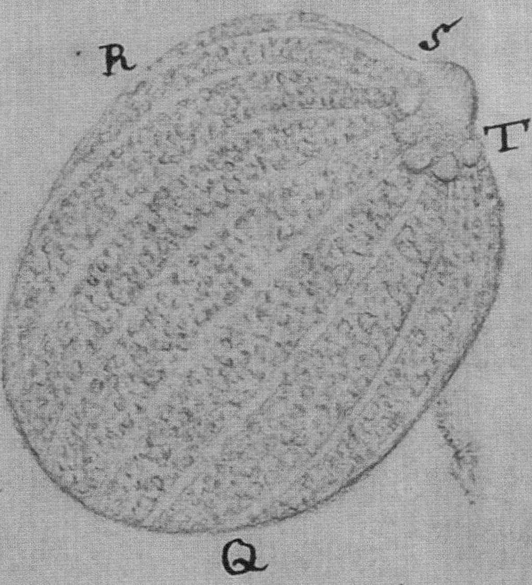

fig

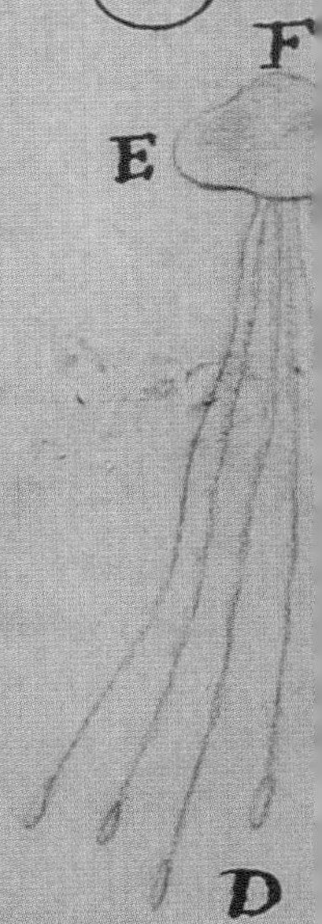

fig:4.

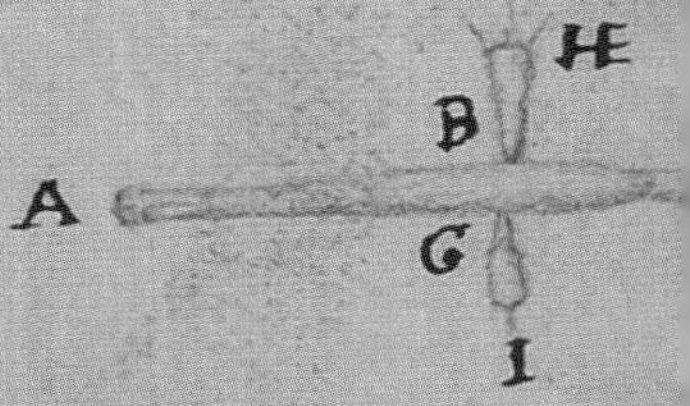

fig:5

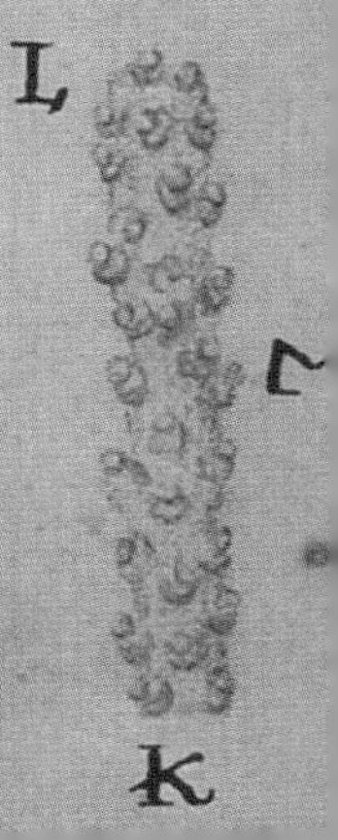

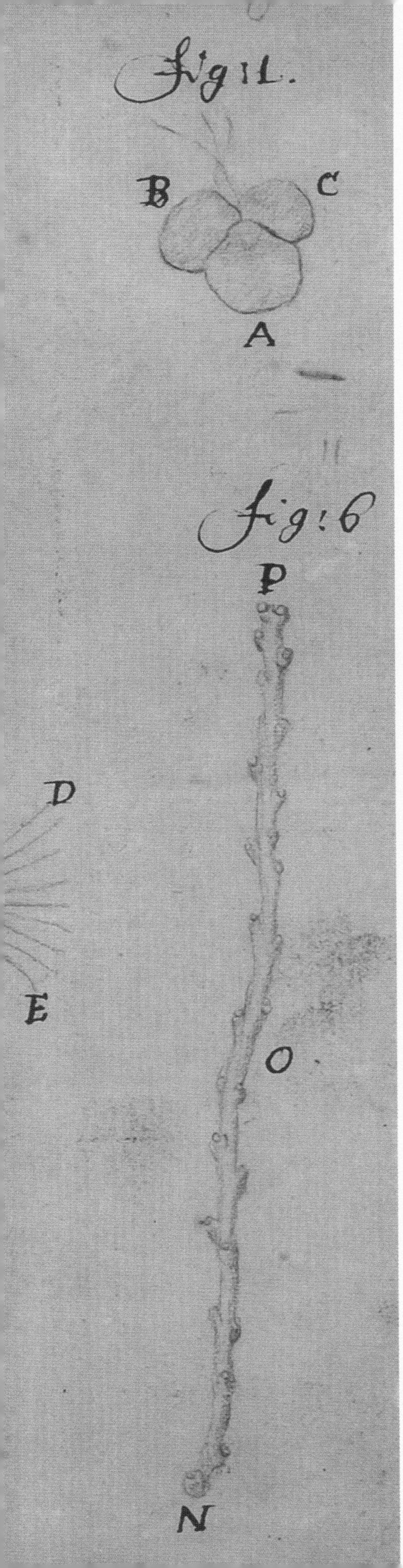

CHAPTER TWO

What makes things happen?

07

The observation of falling objects (1586)

Q Do all objects fall at the same rate?

Intuition tells us that the heavier objects are, the faster they fall. But intuition is often a poor guide in science—and that is why experiments matter so much. For all that Aristotle was a keen observer of the natural world without parallel in ancient Greece, the lack of a tradition for putting ideas to the empirical test led him to assert this intuitive view that indeed the time taken for an object to fall a given distance depends on its mass. A falling leaf, feather, or snowflake, for example, seem after all to verify that principle, although we now recognize that their descent is slowed by air resistance.

During the Middle Ages, Aristotle's natural philosophy was largely endorsed by the church, and it was not until the Renaissance that it became more acceptable to reconsider what he said about how the world works—and to put it to the test.

Aristotle's natural philosophy was teleological, if not indeed tautological: objects behaved in the way they did because they had a natural tendency to do so. Bodies fall to the ground because they seek their natural resting place at the center of the earth; celestial bodies like the stars, Sun, and Moon, remain aloft in the heavens because that is their natural place.

In popular legend, Aristotle's view of objects freely falling under gravity was disproved by Galileo in an experiment from the Leaning Tower of Pisa. In reality, it seems unlikely Galileo ever performed that test. But someone else did, several years before Galileo wrote on the matter. That person was the Flemish engineer Simon Stevin, who used instead the tower of the Nieuwe Kerk in the Dutch city of Delft.

In 1586 Stevin and his friend Jan Cornets de Groot ascended the tower with two heavy lead balls, one ten times the weight of the other. They dropped the balls 30 feet onto a wooden platform below, judging the time of descent from the sound of impact, as well as eyewitness testimony of the fall. As well as could be assessed, the two balls struck the wood at the same moment. Stevin described these observations in a book written that same year.

Seventeenth-century anonymous portrait (oil on cloth) of Flemish engineer Simon Stevin, from the Universitaire Bibliotheken Leiden, the Netherlands.

Detail of the belltower of the Nieuwe Kerk. From Hendrick Vroom's oil on canvas *View of Delft from the Southeast*, 1615, Museum Het Prinsenhof, Delft.

It seems unlikely that even Stevin was the first to question this aspect of Aristotelian physics. As early as the sixth century AD, the Byzantine scholar John Philoponus of the school of Alexandria challenged that picture of free fall, recording that an object twice the weight of another differs imperceptibly in its time of descent. But slight imperfections and inconsistencies in Aristotelian mechanics were minor aberrations. Galileo's studies suggested that a more profound overhaul was needed.

Galileo was born in Pisa, and entered the city's university in 1580 as a student of medicine. After graduating, he was given a lecturing post at the university in 1589, where he stayed until moving to Padua in 1592. The story of his Tower of Pisa experiment comes from Vincenzo Viviani, who became his assistant and devotee toward the end of Galileo's life and subsequently wrote a biography that was, in the manner of its time, both hagiographical and of questionable reliability. Viviani says that Galileo performed his experiment, presumably sometime during his spell as a lecturer in Pisa, "in the presence of other

SIMON STEVIN | 1548–1620

A mathematician and engineer, probably born in Bruges, Simon Stevin served in the court of William, Prince of Orange, who led a revolt against Spanish rule in the Low Countries. After the Prince's assassination in 1584, Stevin became the principal advisor to his son and heir Maurice, designing military fortifications and public works such as improvements to Dutch windmills. At Maurice's request he founded an engineering school at the University of Leiden.

See also: Experiment 2, Direct demonstration of the rotation of the Earth, 1851 (page 16); Experiment 8, Deducing the law of acceleration in free fall, 1604 (page 38).

teachers and philosophers and all the students." There is no other record of the event—not even in Galileo's own works, which refer only to tests of free fall made with a cannonball and musketball. We might reasonably suspect that Galileo did conduct experimental tests of free fall, but a dose of scepticism is wise regarding whether any of them were from the Leaning Tower. It's likely Viviani was seeking to engage his readers with a good story.

But if the Pisa experiment is a myth, Galileo perceived nonetheless that this was about much more than how objects fall: he believed the cosmos is governed by general laws and principles that apply to bodies universally. If Aristotle was wrong about free fall, that crack spread further: an entire new science of motion was needed. This was what Galileo presented in his *Discourses and Mathematical Demonstrations Relating to Two New Sciences* in 1638, four years before his death.

Stevin's tower test is a perfect example of how mere observation morphs into experiment. It involved testing a falsifiable hypothesis (say, "time of fall depends on mass") by manipulating objects in a controlled way, as opposed to simply observing a natural phenomenon. There is measurement and collection of data: the sound of impact quantifies the descent time. There is a variable that the experimenter changes—the mass of the balls—to see if there is a difference in outcome. There is a need for the careful experimenter to consider sources of error: how precisely coincident is the time of release of the balls, say? Yet the test remains close to the older notion of "experience": we simply look for this or that outcome, which relates to the interpretation in a straightforward way. There is no specialized apparatus involved, and no real quantification

Copper engraving of Jan Cornets de Groot. From *Illustrium Hollandiae & VVestfrisiae ordinum alma Academia Leidensis*, 1615, Lugduni Batavorum: At James Marcus, & Justum à Colster, Booksellers, Getty Research Institute, Los Angeles.

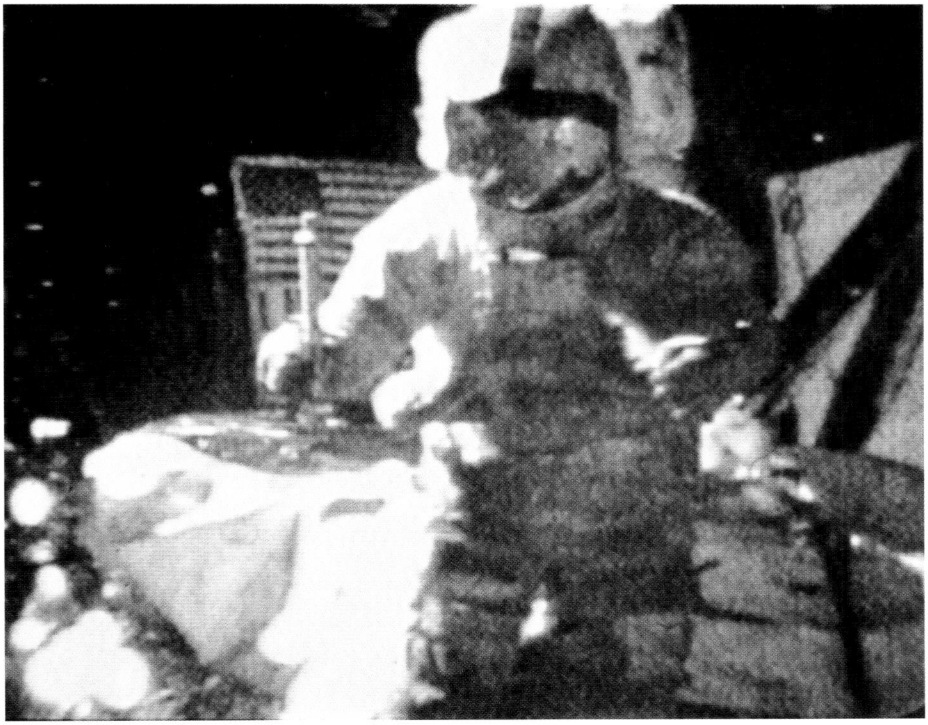

The Hammer-Feather Drop experiment: film still of Commander David Scott recreating Galileo's experiment during a Moon Walk on the Apollo 15 Mission, August 2, 1971, which shows that objects fall at the same rate in a vacuum. Image courtesy of NASA.

of data. At the core of the matter, though, was the new spirit of the age: do not take the old authorities on trust, but look for yourself.

Galileo was the first to adduce (if not in quite these terms) the role of air resistance in retarding the fall of an object like a feather. If there were no air—if they fall in a vacuum—a pebble and a feather will fall at the same rate. That was demonstrated in 1969 when Apollo 15 astronaut David Scott dropped a feather and a geological hammer on the Moon. Scott played to the Pisa myth, telling viewers back on Earth that he was demonstrating "a rather significant discovery about falling objects in gravity fields" made by "a gentleman named Galileo." Indeed, both hammer and feather hit the lunar surface at the same moment—"proving," said Scott, that "Mr Galileo was correct in his findings." This was not really an experiment as such, since the result was never in any doubt; it was simply a demonstration in the manner of a school lesson. Yet, the historical shortcomings notwithstanding, it a heck of a way to make a scientific point to a global audience.

Deducing the law of acceleration in free fall (1604)

Q How does the speed of falling depend on time?

It was one thing to know that all objects take an equal time to fall in gravity (if they are not significantly perturbed by air resistance). But how exactly did they fall? Seemingly based on observations of bodies falling through water, Aristotle asserted that they fall at a steady speed that depends on their weight. But the descent in air was too rapid to judge that by eye. If only there was some way to slow it down.

In the early 1600s, Galileo saw how that might be simply done. A smooth ball placed on a slightly sloping tabletop begins to roll. If the slope is steeper, the ball gathers speed more quickly. Increasing the slope until it is vertical brings the motion steadily closer to perfect free fall. So Galileo reasoned that a ball rolling down such an "inclined plane" is a slowed-down version of free fall that would enable him to make measurements.

The question was how the distance traveled (call it s) depends on the time elapsed (t). If the ball rolls at constant velocity, the two are proportional: the speed is then just the ratio of distance to time. Galileo began experimenting with inclined planes in 1602, and two years later he had improved the method enough to deduce the mathematical relationship between s and t. He describes the apparatus in the 1638 *Two New Sciences*. In a wooden beam about 28 feet long, a groove was cut in the edge and covered in smooth vellum, down which a bronze ball rolled when the beam was tilted at an angle. To measure the time taken to reach the bottom after being released from various points along the beam, Galileo used a water clock in which water flowed at a constant rate through a pipe. If the pipe could be opened and closed accurately enough, the amount of water accumulated was proportional to the time passed. Acknowledging the potential for error in this technique, Galileo repeated each experimental run many times—"a full hundred," he claimed.

By this means, he deduced that the ball did not roll at constant speed after all, as Aristotle claimed, but gradually picked up speed: it accelerated. So the relationship between s and t was not one of simple proportionality; instead, s increased in proportion to the *square* of the time elapsed. As students learn to write it today, $s = \frac{1}{2}at^2$, where a is the acceleration. Thus, motion and mechanics were best described not in qualitative language but in mathematics, which Galileo famously declared to be the true language of nature.

Galileo was the first to identify acceleration as a quantity in the theory of mechanics. A body accelerates if a force acts on it—in this case the force of gravity (and also the smaller retarding force of friction as the ball rolls). Galileo deduced that a body on which no forces act does not change speed: if it is already moving, it continues to do so at the same speed, but if it is at rest (a speed of zero), then it remains so. Isaac Newton was later to express this as his first law of motion, and added to it a second law relating acceleration to the force producing it: the force is equal to the mass of the body multiplied by its acceleration a.

Galileo's inclined plane is one of the first instruments designed solely for quantitative experimental science. Previously, natural philosophers tended to use the resources they had to hand—sticks and rods, darkened rooms, prisms and jars—to investigate how nature works. But Galileo's was a genuine scientific instrument constructed with a specific aim in mind. The experiment marked the beginning of a century

Ottavio Mario Leoni's portrait of Galileo Galilei, 1624, Disegni Vol. H, Biblioteca Marucelliana, Florence, Italy.

in which specialized (often costly) scientific instruments became commonplace. Their use increasingly distinguished the "expert" (later sometimes called the virtuoso) from the mere amateur dabbler.

Still we should not suppose that Galileo was doing science in the same sense as scientists today. His method was poised between the older practice of starting with axioms and making logical deductions, and the modern way of formulating and testing hypotheses. As historian of science Domenico Bertoloni Meli says, "he formulated the science of motion as a mathematical construction and then used the experiment only at a later stage to show that the science he had formulated corresponded to nature's behaviour." Even if it did not, Galileo would argue that the science remained valid as a mathematical exercise.

Because of its importance in the history of experimental science, the inclined-plane experiment has received intense scrutiny. One concern was whether Galileo could really have

GALILEO GALILEI | 1564–1642

Born in Pisa, Galileo achieved fame in 1610 with his telescope observations of the rugged Moon and the four major moons of Jupiter—both signs that the heavens did not have the simple perfection attributed by Aristotle. His publication of 1632, *Dialogue of the Two Great World Systems*, offended Pope Urban VIII (previously on friendly terms with Galileo) as much for its mocking tone as for its advocacy of the Copernican sun-centered model of the cosmos. After standing trial and recanting in 1633, he spent the remainder of his days under house arrest in Arcetri, near Florence.

No historical scientist, besides perhaps Newton and Darwin, excite wider admiration among today's practitioners than Galileo. Due to his careful, qualitative experimental work and observation, expressed in the form of mathematical laws, many physicists (who are apt to see this as the epitome of real science) consider him to be the father of modern science. To historians, he is a more contested figure. Undoubtedly brilliant and original, he could also be arrogant and unduly confident in his conclusions (he was wrong about tides and comets, and could not accept Johann Kepler's elliptical planetary orbits). The treatment (albeit not torture) that he received from the Catholic Church for his avowal of the heliocentric theory was a shameful triumph of dogma over scientific inquiry—but a less argumentative and provocative person might well have managed to avoid that fate. Whatever Galileo's personal complexities, however, the careful experiments and observations he conducted marked a significant stage in the evolution of modern science.

Wood and brass reconstruction (214 inches long) from the early nineteenth century of Galileo's inclined plane, Museo Galileo, Florence, Italy.

achieved reliable results with the rather crude methods at his disposal for measuring time. It takes only seconds for the ball to descend, so there is scope for significant error in determining exactly when it starts and ends. With this in mind, French philosopher of science Alexander Koyré was bitingly dismissive of the whole story, saying in 1953 "it is obvious that the Galilean experiments are completely worthless: the very perfection of their results is a rigorous proof of their incorrection." But Koyré's doubts were themselves questioned in 1961 when Thomas Settle, a student of the history of science at Cornell University, showed using cheap homemade equipment that with practice he could obtain data perfectly good enough to verify Galileo's law of acceleration. Koyré's argument from personal incredulity was not enough; it is now common for historians of science to make historical reconstructions of experiments with resources available at the time to see if the findings were plausible. Settle's work also showed the importance of getting a feel for one's apparatus before it could be used with confidence.

What is more, in 1972 historian Stillman Drake concluded from a close inspection of Galileo's scrappy "lab book" jottings that Galileo might have also used another timing method, which involved inserting moveable gut frets into the inclined

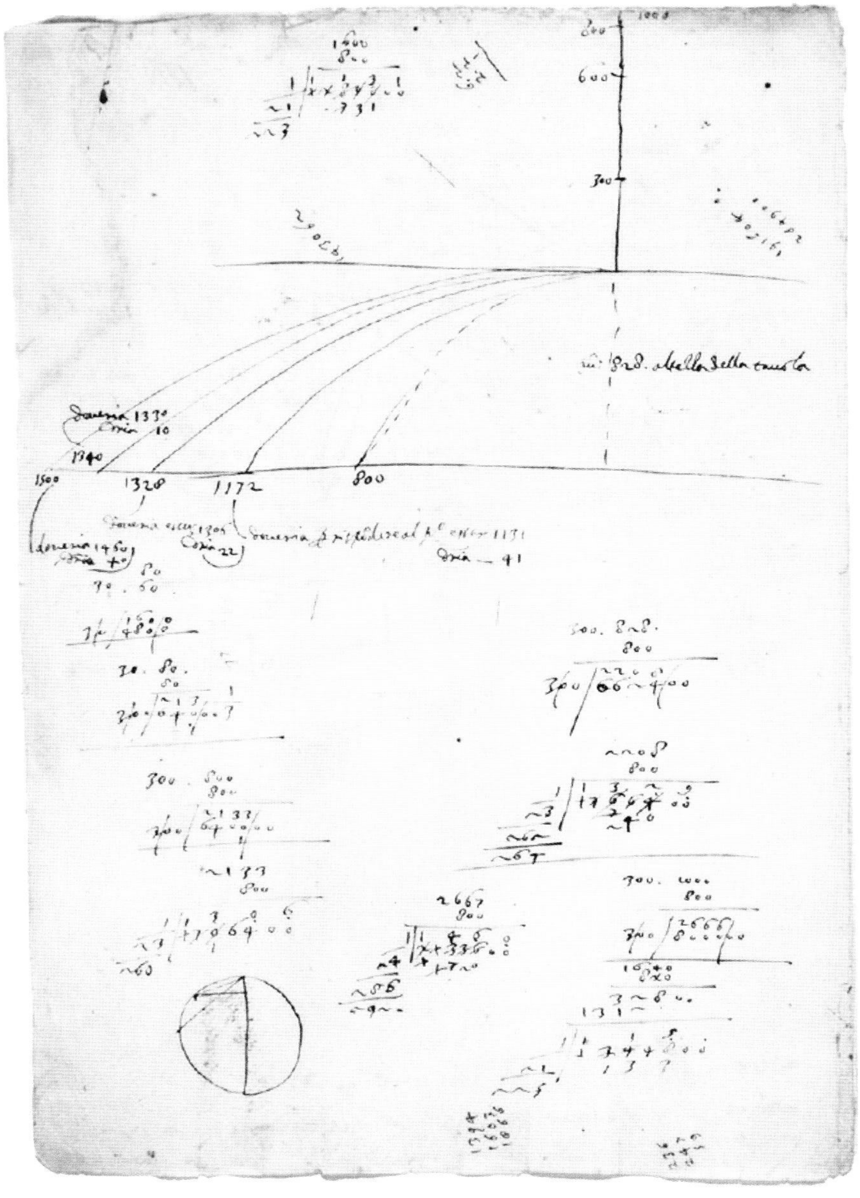

Sketch of projectile motion. From Galileo's notebook, *Discorsi*, 1604, Galileo's Manuscript 72, Biblioteca Nazionale Centrale, Florence, Italy.

plane so that the rolling ball would create an audible click as it passed over. By relying on the good sense of a steady beat that his musical training would have instilled, Galileo might then have moved the frets until the ball's passage produced a regular series of clicks, and deduced the acceleration law from the distances between the frets as the ball rolled for an equal time. Drake suggested that Galileo had not recorded this timing method, perhaps for fear that it might sound foolish—requiring that he establish a regular rhythm by, say, singing a song.

The effects of air pressure (1643/1648)

Q What keeps a liquid in an inverted tube, and how does this change with elevation?

Moving water by pumping was one of the most useful technologies of the ancient world. By this means water could be extracted from deep wells and flooded mines, and distributed for irrigation. In ancient Egypt the main water-lifting technologies were the shaduf, a simple weight-lifted bucket, and the waterwheel. Archimedes, the great engineer of Hellenist Greece in the third century BC, is commonly credited with the invention of the screw pump: a device that raises water in a rotating helical channel. His contemporary Ctesibius of Alexandria devised the kind of pump with which we are most familiar today, in which pistons moving inside cylinders draw up the water to fill the empty space created when the piston is withdrawn. Ctesibius used such a device to pump air for the first known pipe organ.

Piston pumps seem to illustrate the principle articulated by Aristotle that "nature abhors a vacuum": in short, air or water will rush in to fill an empty void. For Aristotle, this precept reflects a teleological universe: a void contravenes what is natural. But it is one thing to say that nature will try to avoid a vacuum, another to suppose that such a void is impossible. Could experiments compel nature to, so to speak, accept something against her will?

Such esoteric questions could seem a long way from the practical limitations of pumps discovered by experience. To the frustration of mining engineers seeking to clear water from deep shafts, water pumps proved unable to raise water more than about 34 feet. In his mining treatise of 1556, *De re metallica*, the German engineer Georgius Agricola shows the cumbersome solution: water had to be lifted from deep down in several stages.

Galileo was aware of this problem, and he suspected that it touched on the question of whether a vacuum is possible. If a pump lifts a column of water, the weight of the water pulls it back down. Galileo speculated in his *Two New Sciences* (1638) that for a water column taller than 34 feet, the weight of the water will break the column and open up a void, preventing the column from rising.

In the early 1640s, the Italian mathematician Gasparo Berti in Rome tested the idea using a vertical lead pipe 36 feet tall, capped with a glass dome. If the pipe was filled with water, it ran out until the level reached that limiting height of 34 feet, with the globe being empty. What was in there? Air—or nothing?

Another Italian scholar, Evangelista Torricelli, saw this experiment in a different way. He wondered what kept the rest of the water in Berti's pipe at all, and concluded that it was held there by the downward pressure exerted by the atmosphere—an "ocean of air"—on the water that flowed out at the bottom into the container in which the pipe stood.

In 1643 Torricelli devised an equivalent but much more convenient experiment in which a glass tube, sealed at one end, was filled with liquid mercury and inverted in a dish, open end downward. Because of the greater weight of mercury, the column now had a limiting height of less than 3 feet. Air pressure on the surface of the mercury-filled dish, and not nature's abhorrence of a vacuum, holds it in place, Torricelli said.

Again, the column falls to leave an open space at the top, which became known as the Torricellian gap. Torricelli thought that it

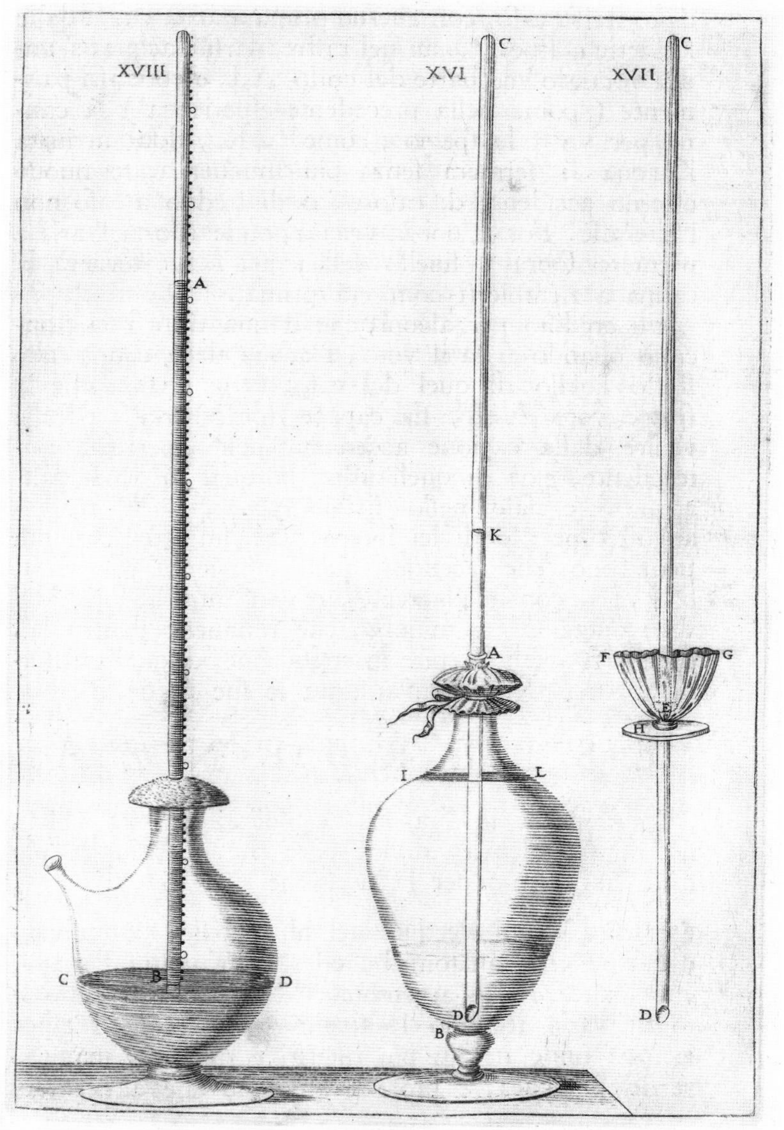

Torricelli mercury barometers. From the Accademia del cimento's *Saggi di naturali esperienze*, Florence: Per Giuseppe Cocchini all'Insegna della Stella, 1666, Plate LXI, Rare Book and Special Collections Division, Library of Congress, Washington DC.

contained a vacuum, but this was widely disputed. He showed that in any event the properties of the supposed vacuum made no difference to the limiting height of the mercury column, for it remained the same if the tube was simply sealed at the top end or if a small glass globe were installed there.

Torricelli's experiment and his interpretation imply that the pressure of air at the Earth's surface is enormous: enough to hold up a 34-foot column of water. It was no news that air itself has weight: Aristotle had said as much, and Galileo appealed to that authority when he repeated the claim in his *Two New Sciences*. But who would

EVANGELISTA TORRICELLI 1608–1647

As a student of the sciences in Rome, Torricelli was taught by the Benedictine monk Benedetto Castelli, who was himself taught by Galileo. When Castelli later sent to Galileo (then living under house arrest at Arcetri) Torricelli's treatise on dynamics and ballistics, Galileo invited the younger man to visit. He did so in late 1641, learning from Galileo in the three months before the latter died. Torricelli subsequently succeeded Galileo as court mathematician to Grand Duke Ferdinando II of Tuscany and professor of mathematics at the University of Pisa, where he continued to work on mechanics, geometry, and optics until his early death, probably from typhoid.

See also: Experiment 10, Vacuums and the elasticity of air, ca. 1659–1660 (page 48); Experiment 15, The discovery of oxygen, 1774–1780s (page 70).

have imagined that this pressure is so enormous, when we have so little experience of it? Torricelli was not surprised, however, because he estimated that the atmosphere is about 50 miles thick and therefore a massive column of air stands above every spot on the ground. Galileo estimated the weight of air experimentally, putting it at around one four-hundredth of the equivalent volume of water—a figure that Torricelli took on trust, although it is about twice the real value (forcing one to wonder if Galileo truly conducted that experiment). At any rate, the immense force of air pressure was famously demonstrated in 1654 by the German philosopher Otto von Guericke, who used an air pump to evacuate the space enclosed by two close-fitting copper cylinders, 20 inches in diameter, and showed that two teams of horses were unable to pull the hemispheres apart. This eye-catching demonstration of atmospheric pressure attracted wide interest at the time, as well as showing the potential of air pumps to produce a vacuum.

If the level to which the mercury column falls is determined by the pressure of the air, this pressure should be reduced on higher ground, where the mass of air pressing down from above is lessened. Torricelli seemed to recognize this, saying that "The weight referred to by Galileo applies to air in very low places, where men and animals live, whereas that on the tops of high mountains begins to be distinctly rare and of much less weight than four-hundredth part of the weight of water."

Evangelista Torricelli. From the frontispiece of *Lezioni accademiche d'Evangelista Torricelli*, Florence: Nella Stamp. di S.A.R. Per Jacopo Guiducci, e Santi Franchi, 1715, Rare Book and Special Collections Division, Library of Congress, Washington DC.

Gaspar Schott, *Technica curiosa, sive mirabilia artis*, Nuremberg: Johann Andreas Endter (printed by Jobst Hertz, Herbipolus, 1664), Wellcome Collection, London.

In 1648 the French philosopher Blaise Pascal put this idea to the test by conducting the mercury-tube experiment at a considerable elevation. He had the apparatus carried up the mountain of the Puy-de-Dôme in the Massif Central, whereupon the mercury level fell by several inches. Pascal did not do the 3,000-foot mountain trek himself, but persuaded his brother-in-law Florin Périer, who lived in Clermont in that region, to do so. The diligent Périer claimed that he repeated the measurement at five places on the summit, all showing a mercury drop of 3½ inches, as well as checking that the mercury fell less at stops farther down the mountain. He also tried the experiment by climbing up the tower of the cathedral in Clermont, which produced a perceptible fall in the mercury. Some historians have disputed the authenticity of Périer's account in his letter to Torricelli—but if it is true, it is sometimes regarded as the first formal experiment in being motivated by a clear hypothesis, designed, measured, and repeated. The great proto-scientist Robert Boyle claimed that Pascal and Périer's test marked the validation of the experimental philosophy.

Torricelli's inverted mercury tube is, in fact, a barometer: the height of the mercury column measures the local air pressure. This changes not only with elevation but also due to changes in atmospheric pressure connected to the weather, which is why barometers became used for weather forecasting.

The impact of new techniques

We discovered earlier that theories generally precede rather than flow from experiment (see page 24). If that's so, how does science ever develop new concepts, rather than just validating or falsifying those it has already? That's a complicated question, but one answer surely lies with the development of new techniques. From the microscope and the telescope to the observation of the universe at wavelengths the eye cannot register, our view of the world has repeatedly been transformed by the invention of new techniques for studying it.

It is sometimes suggested that new scientific techniques and instruments allow us to detect phenomena that we could not detect before. Occasionally, this is true. An experimenter might say, for example, "I know that there are certain kinds of molecule moving around in these cells, but I can't see them—so I shall devise a method that reveals them." But new techniques may also disclose phenomena we never knew existed, and to explain for which we therefore have no pre-existing theoretical framework. That was true, for instance, with C. T. R. Wilson's cloud chamber (see page 121), which revealed the existence of cosmic rays. The development of photography allowed Henri Becquerel to discover radioactivity (see page 88). And, most shockingly perhaps, the early microscopes showed that there are living organisms too small for the naked eye to discern (see page 170). Such phenomena are, as it were, launched into the world theory-free, and what often follows is a gloriously and (one hopes) productively messy free-for-all. It's not too much to say that new techniques can then change reality, if by that we mean all things we know or believe to exist. Some experiments, says philosopher of science Ian Hacking, don't merely discover but *create* new phenomena. They can transform hypothetical objects, such as the atom, the neutrino, and the Higgs boson (see pages 104, 128, and 130, respectively) into real ones.

Traditional histories of science often celebrate breakthroughs in thinking: Newton perceiving the unity of gravitational phenomena; Lavoisier realizing that the oxygen hypothesis explains combustion better than phlogiston (see page 68); and Charles Darwin conceiving of the theory of evolution by natural selection. It's an understandable bias, for these are the ideas that reshape our understanding of the world. But the impact of new instruments and techniques on the advance of science is incalculable. The introduction of X-ray crystallography (see page 158) is surely among the most profound, for it allowed us for the first time to peer into the atomic realm and see how atoms are arranged—with implications for chemistry, biology, metallurgy, mineralogy, and more. Some new techniques provide answers only to narrow questions, but a method like this reverberates through the sciences. Their importance is reflected in the scientific Nobel prizes, many of which have been awarded for the development of a new technique rather than for new discoveries or theories.

Scientists are only as good as their tools

With a new method come new problems, however. When a scientist uses tried-and-trusted instruments like a microscope or galvanometer, they assume that the device, provided it is not faulty, will do what they think it should, for that has been shown many times before. But claims made for an entirely new instrument or technique can't be afforded such trust. New instruments haven't yet been perfected, and are often working at the limits of what can be detected—how could we be sure, for example, that the rise and fall of electrical current measured by a scanning tunneling microscope (see page 116)

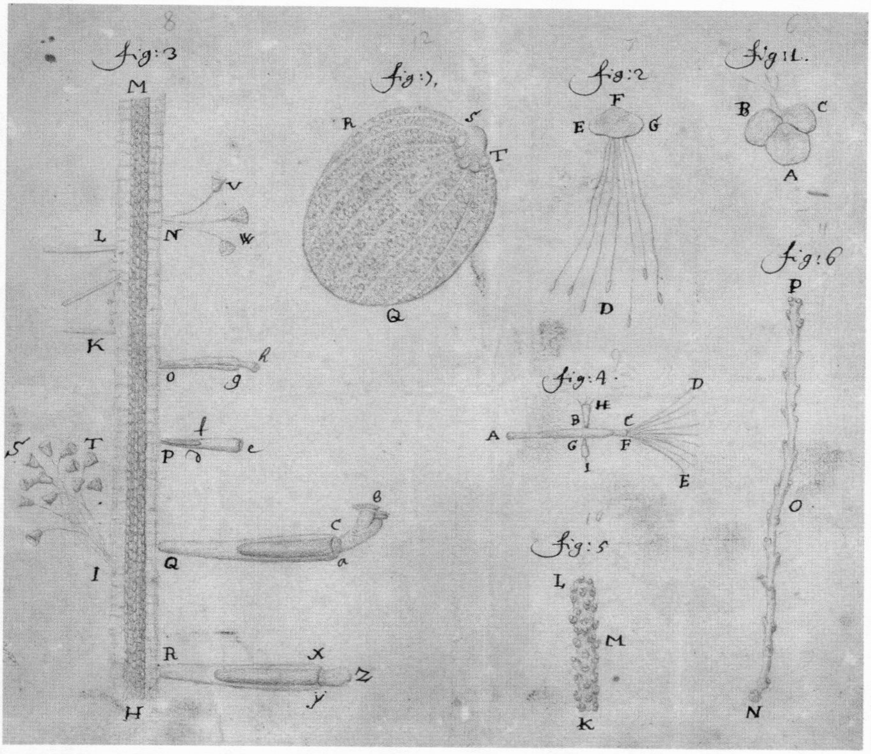

Duckweed and freshwater microorganisms (*diertjes*, or "animalcules"). Illustration from Antonie van Leeuwenhoek's *Letter to the Royal Society*, December 25, 1702, The Royal Society, London.

was really due to individual atoms, and not just random noise? New techniques have created plenty of false alarms: yes, photographic plates revealed X-rays and radioactivity, but also several other new types of "ray" that turned out to be figments of the imagination. It is only natural and proper, then, that new experimental techniques and tools are often met with some scepticism. An account in 1610 of an early demonstration of the telescope by Galileo in Bologna (an instrument he did not invent, though he was its most famous early adopter) tells how he slunk away dejected after failing to persuade his learned audience that the device worked. "All acknowledged that the instrument deceived," says the author. "Full of himself, he hawked a fable." Galileo had the last laugh, however, because the short treatise he published that year, *Sidereus Nuncius*, made him a celebrity by reporting the moons of Jupiter and the rugged terrain of our own Moon.

The problem, says historian of science Albert Van Helden, is that until the advent of reliable twentieth-century instruments that recorded on photographic film, telescopic observation was a private act, at which some were better than others. That remains so for many scientific techniques, especially when they are bespoke. To measure the charge on an electron, Robert Millikan (see page 112) had to acquire a feeling for his apparatus. Experiments on biological systems, such as Hilde Mangold's tissue-transplantation studies (see page 192), are notoriously erratic and often require instruction from an expert rather than mere recipe-following. Like a musician, one has to learn how to "play" the instrument. Only then can the tunes be pleasing, well performed, and even revelatory.

Vacuums and the elasticity of air (ca. 1659-1660)

Q How "springy" is air?

Experiments involving vacuums showed how unsettling the experimental philosophy could be to our intuitions about the world. That empty space is filled with air was just common sense; the notion of truly empty space containing nothing at all seemed to many philosophers to be disturbing and unnatural. Even if it could be created, surely nature did its best to destroy it?

The Anglo-Irish scholar Robert Boyle didn't see it that way. He wanted to study vacuums not in the ad-hoc manner of von Guernicke or Torricelli (see page 46) but systematically, in a way that gave him control over the relative proportions of air and of nothingness.

The youngest son of the first Earl of Cork in Ireland, Boyle joined a group of like-minded natural philosophers in Oxford in the 1650s who were attracted by the new experimental philosophy. In 1657 Boyle heard about the pump that von Guernicke had used to demonstrate air pressure, and desired such a device to conduct his own studies. Being a nobleman of means, he hired the inventor and experimenter Robert Hooke—a man of much more humble origins—to build him an improved version, which was called the Pneumatical Engine.

In place of von Guernicke's copper hemispheres, Boyle and Hooke used a large spherical chamber made of glass, with a lid in the top through which items (including hapless small animals) could be placed in the vacuum chamber and observed. The bottom of the chamber had an outlet connected to a hand-cranked pump that extracted the air—a difficult process, because on each stroke of the crank, a valve had to be opened and then closed to prevent air being sucked back in. As the chamber became more and more empty of air, it was ever harder to draw the last of it out. Keeping the seals airtight with putty was a battle that the two men were constantly fighting: the vacuums this device could deliver were far from perfect, and the pair needed time and assistance to get it working at all.

All the same, Boyle's air pump allowed him to conduct experiments that no one else could. It was an exclusive and costly piece of kit, often now compared to the huge particle accelerators that enable today's physicists to conduct studies accessible to them alone. That exclusivity poses a problem for experimental science: how trustworthy are your results if no one else has the means to independently check them?

In the late 1650s, Boyle and Hooke carried out a wide range of investigations into the nature of air and the consequences of its absence. They studied whether sound could travel through a

ROBERT BOYLE | 1627–1691

Robert Boyle was born in Lismore, Ireland. Alhough often regarded as an early chemist (his book *The Sceptical Chymist* clarified the notion of an element), he was deeply curious about all the sciences and was one of the founders and key thinkers of the Royal Society of London.

See also: Experiment 9, The effects of air pressure, 1643/1648 (page 42); Experiment 15, The discovery of oxygen, 1774–1780s (page 70).

Joseph Wright of Derby's oil on canvas *An Experiment on a Bird in the Air Pump* (recreating one of Boyle's experiments), 1768, National Gallery, London.

vacuum, concluding that it couldn't (at least for the ticking of one of Hooke's watches). They showed that the force of magnetism still acted in a vacuum: a compass needle would be attracted to a magnet inside the empty chamber. Birds and mice became very sick in a vacuum—such experiments were depicted a hundred years later in Joseph Wright of Derby's dramatic 1768 painting *An Experiment on a Bird in the Air Pump*, which captured the excitement and revelation as well as the moral ambiguity of the dawning Age of Enlightenment. Boyle described the fruits of his studies in his 1659 book *New Experiments Physico-Mechanicall...*, shortly before he moved to London and became one of the founders of the Royal Society, the key locus of the experimental philosophy in England.

One particular issue that exercised Boyle (and Hooke, who had ambitions of his own) was the nature of air itself. We are familiar now with the idea that air, and gases generally, are fluids that simply expand to fill the space available. If some air is pumped out of a chamber, that which

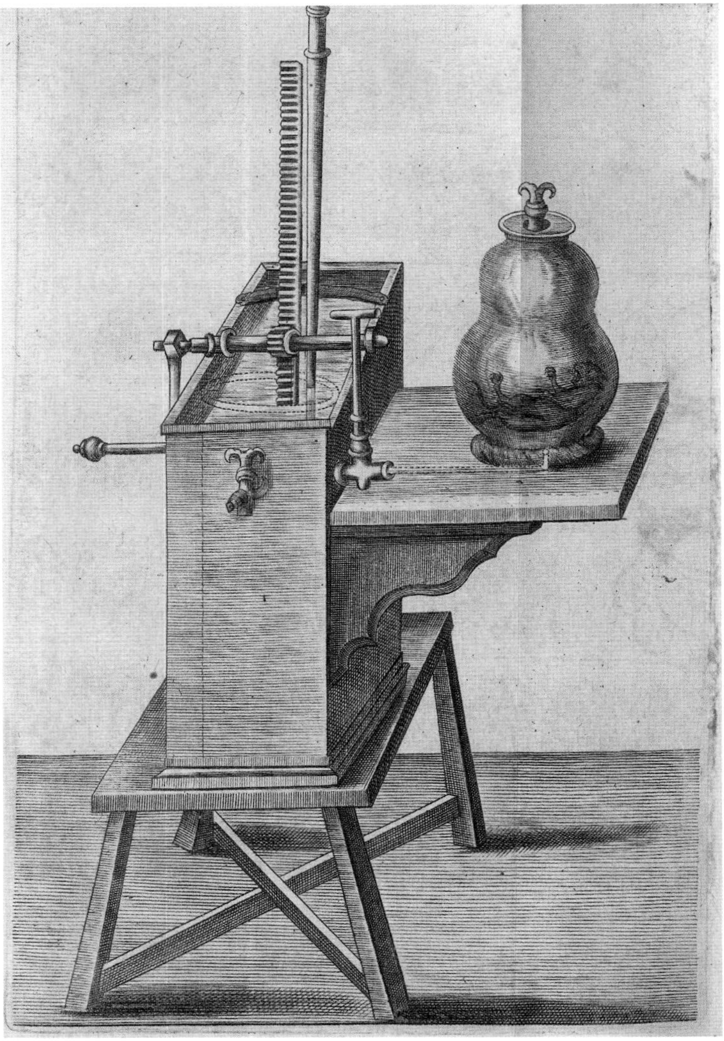

Experiment demonstrating that an animal dies in a vacuum. From Robert Boyle's *A Continuation of New Experiments, Physico-Mechanicall, Touching the Spring and Weight of the Air*, Oxford: H. Hall, for R. Davis, 1669–82, Vol. 1, Plate I, Rare Book and Special Collections Division, Library of Congress, Washington DC.

remains gets more spread out. Boyle didn't quite picture it this way, though. He and his contemporaries thought of air as a kind of elastic substance with a certain "springiness": Boyle compared it with sheep's wool, which will shrink when compressed but then spring back again when it is released.

Boyle measured this "elasticity" of air using simpler apparatus than the air pump. It was a model of minimalist elegance: a J-shaped glass tube closed at the shorter end, into the open end of which Boyle and Hooke poured mercury to trap a bubble. They tipped the tube initially to open up a passage between the bubble and the air outside so that their pressures could equalize. Then they added more mercury to the open end, under the extra weight of which the trapped bubble contracted. The increased pressure ("spring") with which the bubble pushed back on the column of mercury could be deduced from the weight of

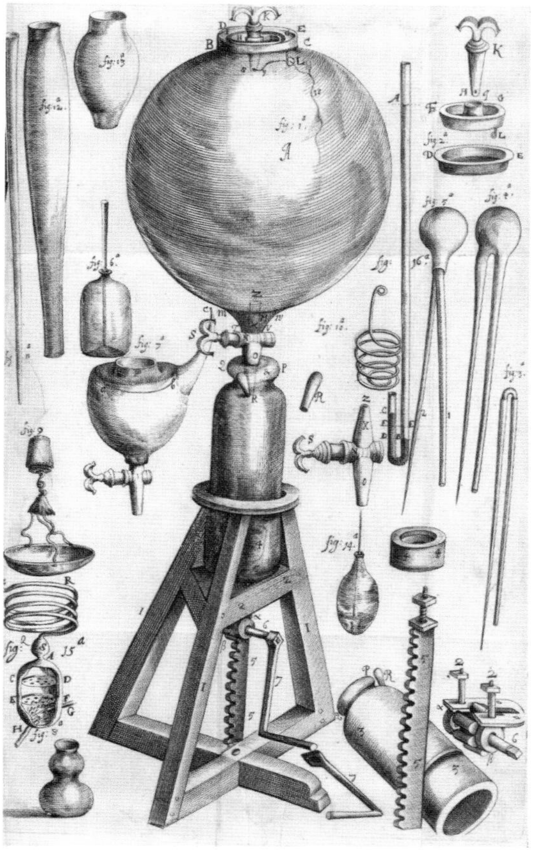

Copperplate engraving of a pneumatical engine and components. From Robert Boyle's *New Experiments, Physico-Mechanicall, Touching the Spring of the Air, and its Effects,* Oxford: H. Hall, for T. Robinson, 1660, Science History Institute, Philadelphia.

the mercury added, and in this way Boyle and Hooke could see how the pressure of the air depended on the volume it occupied.

The experiment wasn't without mishaps: the first tube they used broke under the weight of the mercury (spilled mercury is both messy and hazardous), so they repeated it with a sturdier tube housed in a wooden box.

Boyle reported these studies in a 1662 book *A Defence of the Doctrine Touching the Spring and Weight of the Air*. Here he pointed out that when air expands, its pressure decreases in inverse proportion to its volume: if the volume doubles, the pressure halves. This is now known as Boyle's law and is one of the so-called "gas laws" that relate the pressure, volume (or density), and temperature of a gas. Boyle was a little lucky to get sole credit, given the help he received from Hooke. His assistant has got a law of his own now—Hooke's law of springs—but he also discovered an unfortunate law of scientific experiments: technicians rarely get the acknowledgment they deserve.

Anonymous oil on canvas *The Hon. Robert Boyle, experimental philosopher*, ca. 1690s, Wellcome Collection, London.

The origin of heat (1847)

Q Is heat really just atoms in motion?

Fire was one of the canonical quartet of elements of ancient Greece that featured in Aristotle's scheme: a substance in its own right. But by the eighteenth century fire was understood to be a chemical process that produced heat; it was heat itself that then became regarded as elemental. Some considered heat a sort of fluid that passed through tiny spaces or pores in other substances, which is how it was considered to spread. In his list of elements in 1789 the French chemist Antoine Lavoisier listed "caloric," imagined to be the fluid substance of heat.

The caloric theory of heat was accepted by most scientists in the early nineteenth century, but it was not the only explanation. If heat was indeed due to the fluid caloric, a substance could only contain a finite amount of it. But in 1797 the American physicist Benjamin Thompson carried out an experiment that threw this idea into question. As an official in the Bavarian army (in which role he acquired the title Count Rumford) Thompson noted that the boring of cannon barrels produced a great deal of frictional heat. He showed that boring of a barrel immersed in water would generate enough heat to boil the water within a few hours—again and again, seemingly without limit. In a paper presented to the Royal Society in 1798, Thompson proposed that heat was associated instead with motion: the movement of the boring head agitated the particles of the brass barrel.

With the advent of the steam engine and the mechanization of industry in the early nineteenth century, the question of how to generate and use heat most efficiently became pressing. Studies of these problems gave rise to the science of thermodynamics—literally, the motion of heat—over the course of the century. A pioneer of this new discipline was the French engineer Nicolas Léonard Sadi Carnot, who introduced the concept of the cyclic conversion of heat into motion in an engine. Carnot doubted the caloric theory and concurred with Rumford that heat results from the motion of molecules.

James Joule, an engineer from the heart of the Industrial Revolution in Salford, near Manchester, felt the same way in the 1840s. Joule had been taught by John Dalton, the architect of the early atomic theory, and so he was very familiar with explanations of the properties of matter based on the notion of atoms. Joule's studies of the electrical, chemical,and mechanical generation

JAMES JOULE | 1818–1889

James Prescott Joule came from a prosperous family who owned a local brewery in Salford (near Manchester). He was sickly as a child, and his shyness and social awkwardness may have hindered his ability to attract supporters to his ideas. Despite laying some of the foundations of thermodynamics with his demonstration of the conservation of energy, Joule was slow to garner the recognition that his achievements warranted.

See also: Experiment 23, Understanding Brownian motion, 1908 (page 104).

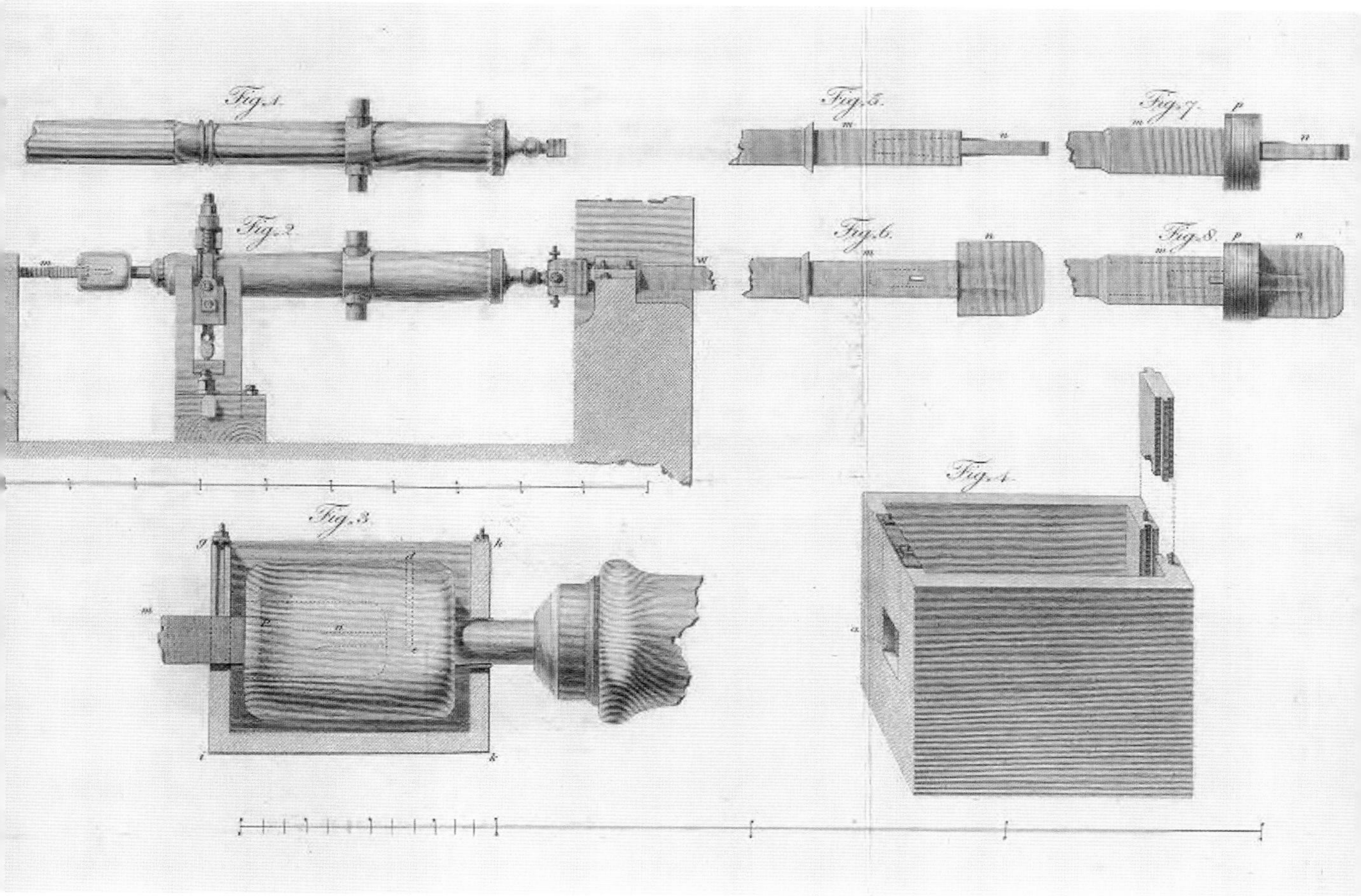

of heat led him to suspect that it was connected to mechanical forces. "Wherever mechanical force is expended, an exact equivalent of heat is always obtained," he wrote in 1850—a comment that can be regarded as a statement of the first law of thermodynamics, which says that energy can neither be created nor destroyed but only interconverted into different forms: in other words, energy is *conserved*.

Joule found a cool reception for his ideas: among those who initially dismissed them were Michael Faraday, the eminent chemist John Herschel, and the physicist William Thomson (Lord Kelvin). It was only when Joule presented his findings at a meeting of the British Association in Oxford in 1847 that his fortunes turned, thanks to a new experiment that seemed to provide support.

Joule made a cylindrical copper container about the diameter of a pudding bowl, inside of which was a paddle wheel that could be turned by strings attached to the axle projecting outside the chamber. These strings passed over pulleys,

Benjamin Thompson, Count of Rumford's "An Inquiry concerning the Source of the Heat which is excited by Friction." From *Philosophical Transactions of the Royal Society of London for the Year MDCCXCVIII* [1798], Part I, Vol. 88, Table IV, Smithsonian Libraries, Washington DC.

George Patten's oil on canvas portrait *James Prescott Joule, with his calorimeter*, 1863, The Manchester Literary and Philosophical Society, UK.

and Joule attached weights to the ends. He could calculate the work done when a weight was lifted, and also the heat this generated inside the cylinder as the turning wheel agitated water or mercury in which it was immersed. The raising of the weight would be repeated quickly many times to generate an appreciable rise in temperature.

After carefully (indeed, tediously) accounting for the various sources of heat loss in his device, Joule showed that the amount of work done by lifting was equivalent to the amount of heat generated. He argued that friction is simply the conversion of mechanical power to heat: in other words, heat and mechanical work are interchangeable. But he was forced to remove that profound conclusion from the paper he published in the *Philosophical Transactions of the Royal Society* in 1850 on the insistence of a sceptical referee: none other than Michael Faraday. "I think," Joule wrote to another physicist soon after, "this view will ultimately be found to be the correct one."

It surely was. Kelvin was eventually converted and began to promote Joule's ideas about the interconversion of heat and motion. He realized that once energy was dissipated as heat, it was hard to recover for doing useful work: that energy was downgraded and "irrecoverably lost." Kelvin realized that, little by little, this must be the fate of all energy in the universe, ending in a cosmic "heat death" where no useful work can be performed and no meaningful change can happen. That was a mighty implication indeed from an experiment with a paddle wheel.

Joule's experiments spelled doom for the caloric theory, which quickly withered. But that battle is in a sense still fought in the units of energy used today. The scientifically respectable unit is named after Joule, but the older "calorie" is still used, especially as a measure of the energy content of food. (Both units are very small, so are multiplied a thousandfold—kilojoule and kilocalorie; kJ and kcal—in dietary contexts.)

It is tempting to suppose that the pattern for important experimental discoveries begins in revelation and ends in acclaim and eminence. Not so for Joule: his arduous experiments were met with scepticism from the scientific establishment, and he struggled both to secure funding and to publish his findings. That, as many scientists today will wryly testify, is at least as likely a trajectory for groundbreaking research.

Philippe Jacques De Loutherbourg's oil on canvas *Coalbrookdale by Night*, 1801, showing the Bedlam Furnaces ironworks, Science Museum, London. The painting shows an iron foundry in Shropshire, symbolizing how the Industrial Revolution was transforming the landscape.

The electric motor and electromagnetic induction (1821/1831)

Q How are electricity and magnetism related, and can we use them to drive machines?

In the early nineteenth century, electricity was one of the greatest puzzles for the natural philosopher—or the "scientist," as the polymathic William Whewell was to name them in 1834. The Italian Alessandro Volta had shown how electricity could be conveniently supplied at will from his "voltaic pile," the first chemical battery. And in 1820 the Danish scientist Hans Christian Ørsted discovered a connection between electricity and magnetism. The discovery seems to have been accidental: Ørsted passed a current from a voltaic pile through an electric wire and noticed that the magnetic needle of a nearby compass twitched. The needle's deflection happened as the current was turned on, and again when it was turned off: it seemed related to a change in current.

News of the finding spread, and a few months later André-Marie Ampère in Paris showed that two current-carrying, parallel wires attract one another like magnets when the currents flow in the same direction, but repel each other when the current is reversed in one of them. It was initially thought that the wires somehow radiate a magnetic influence (a field, as we would now say), much as a hot wire radiates heat and light—but Ørsted deduced that in fact the force of magnetism from the wire encircles it.

Word of these studies reached London, where one of the great experts of the day on electricity was Humphry Davy, director of the Royal Institution. Davy seemingly mentioned the work to his young assistant Michael Faraday. The curious Faraday decided to conduct his own research on the matter, and in September 1821 he set up an experiment to explore the connection between electricity, magnetism, and mechanical motion.

The interaction of the electric current and magnetic field in Ørsted's experiment created a force that moved the compass needle, while Ampère's studies induced movements of the two wires. Now Faraday considered how, in turn, a current-carrying wire might be moved by a magnet. He placed a magnet pole upward in a jar of mercury and suspended a wire from above so that one end dipped into the mercury close to the magnet. He then connected the other end of the wire, and a piece of metal dipped into the mercury, to a battery, forming a circuit that let electricity flow through the wire. This, he found, caused the tip of the wire to circulate around the magnet, executing what Faraday called "electromagnetic rotations."

That was something new, and Faraday quickly wrote up his findings and published them in January 1822 in a paper titled "On some new electromagnetic motions and the theory of electromagnetism" (the latter term had already been introduced to allude to the connection between the two). His triumph was soon dampened by Davy, however, who felt that Faraday had not sufficiently acknowledged the work in this field by Davy's friend William Hyde Wollaston (even though Faraday had mentioned him as well as Ørsted and Ampère). It was an aggressive gesture against a protégé who Davy had seemingly begun to resent as a rival. The relationship between the two men, already strained, never recovered. When Faraday was elected a Fellow of the Royal Society in 1824, the sole dissenting vote against the decision was cast by Davy.

Faraday's experiment is commonly regarded as yielding the first electric motor, although it was not obvious how the rotation of the wire might be

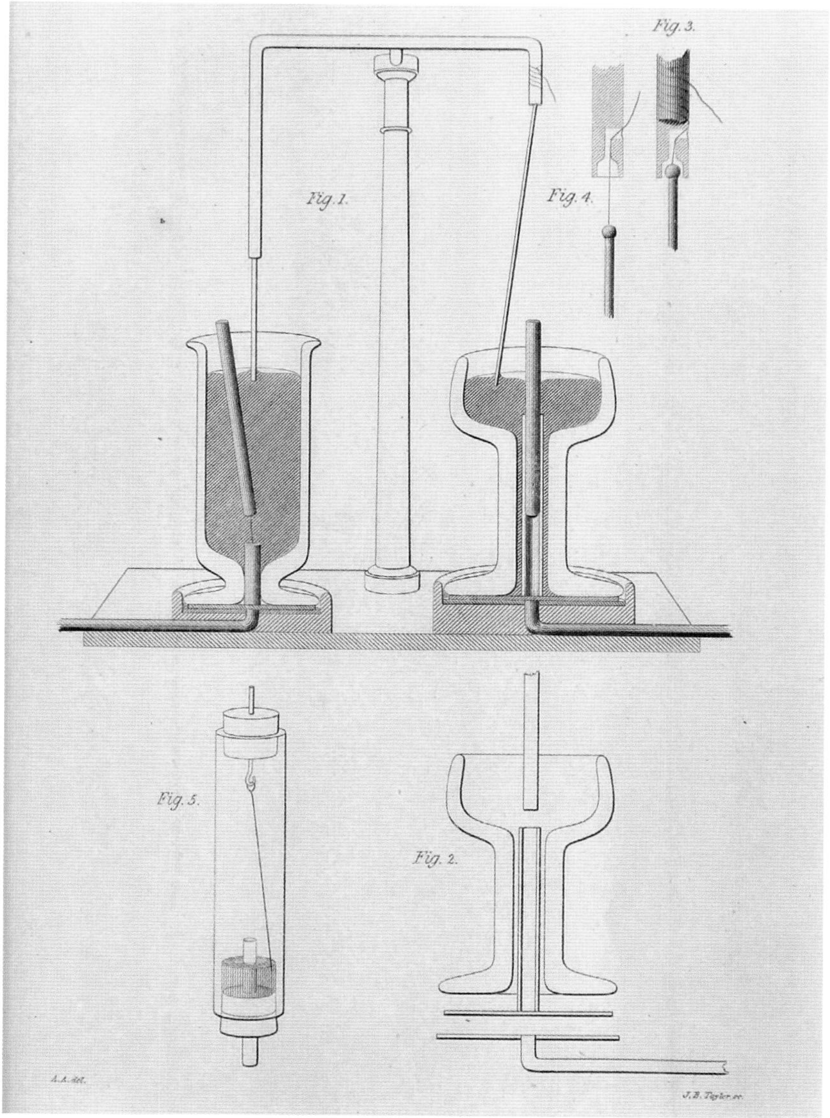

Selection of diagrams showing the apparatus used to convert electrical energy into mechanical rotation, the basis of dynamos, using a bar magnet, beaker of mercury, and current-carrying wire. From Michael Faraday's *Experimental Researches in Electricity*, London: Bernard Quaritch, 1844, Vol. II, Wellcome Collection, London.

usefully harnessed. Within a few months of his results being announced, however, the English scientist William Sturgeon worked out how to convert this circular motion into a practical device using a "commutator": a means of getting the electrical current driving rotation of the device to reverse on each turn as the electrical connecting wires brush against contacts on the rotor itself.

In 1825 Sturgeon managed to demonstrate that an iron rod wrapped around with many coils of insulated copper wire would become magnetized when a current flowed through the coil: this was the first electromagnet. The magnetism could be switched on and off at will, and also increased or decreased by changing the current or the number of coils.

It is not hard to imagine that Faraday himself would have pioneered such discoveries in the 1820s had he been free to pursue his studies. But the demands placed on him at the RI (especially by Davy, who used Faraday more or less as a personal valet) were onerous: in 1825 he was made Director of the Laboratory, a promotion of questionable benefit since it intensified his duties without the compensation of a salary. It was not until the start of the 1830s that Faraday was able to resume experimenting in earnest on electromagnetism—whereupon he made his second major discovery in the field.

If the electromagnet showed that an electric current could induce a magnetic field in an iron bar, might that effect be reversed so that a magnet thus induced could stimulate a current?

MICHAEL FARADAY | 1791–1867

Born in the rural village of Newington Butts, now a part of south London, Michael Faraday was the son of a Cumbrian blacksmith. With such humble origins, he received only a basic education and at fourteen was apprenticed to a bookbinder and bookseller. But he was determined to expand his learning by attending scientific lectures in the city. Given tickets to attend some of Humphry Davy's lectures at the Royal Institution in 1812, Faraday took careful notes, wrote them up in a fair copy (with illustrations), and presented them to Davy. Davy recognized the potential of the young man and in 1813 offered him the post of laboratory assistant at the Royal Institution—later to find to his discomfort that his protégé's acumen and reputation threatened to eclipse his own.

See also: Experiment 16, The discovery of alkali metals by electrolysis, 1807, (page 72); Experiment 42, Animal electricity, 1780–1790 (page 172).

In 1831 Faraday had the idea of curving the iron bar into a ring and wrapping a second coil around it on the opposite side of the coil that induced the ring to become magnetic.

It was no small matter to procure such apparatus. The RI had dedicated artisans who supplied the laboratories, equipped with a furnace and forge for smithying, and Faraday asked the technician Charles Anderson to make him a sturdy iron ring seven-eighths of an inch thick and 6 inches in diameter. He wound it with two coils of copper wire and found that on making and breaking the connections of one coil to a battery, he could detect a current in the other (using a current sensor called a galvanometer)—even though there was no electrical contact between them. This was the first demonstration of a so-called induction ring: the key component of an electrical transformer, which is a ubiquitous device in electrical engineering today.

Faraday reported his results in the first of a series of papers presented to the Royal Society under the umbrella title of "Experimental researches in electricity." The first was read to the society in November 1831, although it took some months before it was published—to Faraday's alarm, for he was worried that the scientists of France, also working hard on the topic, "may get some of my facts in conversation, repeat them & publish in their own name before I am out." Priority in publication was as much a concern for scientists then as it is today.

Faraday continued to report his "Experimental researches" in a series of thirty papers published in *Philosophical Transactions* between 1832 and 1856. They are meticulous accounts of the systematic way in which Faraday investigated various experimental permutations, and testament to the thoroughness with which he recorded his studies in lab notebooks—a discipline impressed on all young researchers today. William Henry Bragg, who became director of the RI's research laboratory in 1923, described what it is about these records that makes them so exemplary:

Faraday was in the habit of describing each experiment, in full and careful detail, on the

Harriet Jane Moore's watercolor *Michael Faraday in his Basement Laboratory* (Albemarle Street, Royal Institution), 1852, The Royal Institution, London.

day on which it was made ... the Diary is far more than a catalogue of results. The reader is able to follow and advance, step by step, to final and fundamental conclusions. He sees the idea forming, its experimental realization, and its employment as a foothold for the next advance.

Faraday's experimental electromagnetism work established the basis of the electrical technologies that burgeoned in the later nineteenth century: the generation, transmission, conversion, and applications of electrical power, from the first generators that drove the electrification of society to the machines that replaced steam engines, and the telegraph that began the age of telecommunications. Faraday also established a connection between electromagnetism and light, which James Clerk Maxwell explained in 1865 when he showed that his own theory of electromagnetism predicted self-sustaining waves of electrical and magnetic fields: the electromagnetic radiation that we know as light itself.

CHAPTER THREE

What is the world made from?

From cataloguing elements to studying the shapes of molecules

Variety from simplicity: this might be regarded as science's guiding principle. The world contains an abundance of substances and phenomena, and the goal is to discover the fundamental source of such profusion. Biology seeks an organizing schema in the theory of evolution, physics in the unification of particles and forces. But the quest is arguably most ancient in chemistry, where the composition and properties of material objects have long been explained by invoking just a few underlying "elementary" substances.

"Proof" that all substances are composed of water (ca. 1648)

Q What is the world made from?

One of the earliest known philosophers of ancient Greece, Thales of Miletus (ca. 624–ca. 548 BC), proposed that all material things in the world are composed of the same fundamental substance: water. This liquid substance could, after all, be transformed into a solid (ice) and a gas (water vapor), and so it seemed to have the potential to adopt all the known states of matter. Thales's idea was disputed by his successors, and throughout the ancient and medical worlds there continued an intense discussion of what the elementary constituents of matter are. The suggestion made by philosopher and poet Empedocles (ca. 490–ca. 434 BC) that there are four elements—earth, air, fire, and water—was popular, but it was by no means universally accepted in the Western world.

However, in the seventeenth century Thales's proposal enjoyed something of a revival, thanks to the work of the Flemish physician Jan Baptista van Helmont. This time, there seemed to be experimental evidence to back it up.

Van Helmont was free to indulge his curiosity about such questions, because he married a wealthy noblewoman and so did not have to rely on earning a living as a doctor. He was an advocate of the new experimental philosophy, and in the family house in Vilvoorde, near Brussels, he conducted many chemical studies, particularly on various "airs" produced in chemical reactions, to which he gave the general name *gas*. At this time, chemistry in the modern sense was just starting to emerge from the older practice of alchemy, and historians often give it the transitional name "chymistry."

In the writings of the Swiss alchemist Paracelsus (1493–1541), van Helmont found inspiration in the idea that life itself might be explained as a chemical process. Like Paracelsus, he believed that life is governed by a kind of internal chemical agency called the archeus. But while this archeus produces our own flesh and blood from the food we ingest, where does the substance of plants come from?

To find out, van Helmont conducted an experiment first described in the fifteenth century

Wood engraving of a willow tree from the title page of Jan Van Helmont's *Ortus medicinae*, 1648, National Library of Rome.

JAN BAPTISTA VAN HELMONT 1580–1644

Like many natural philosophers of his time, Jan Baptista van Helmont, who was born in Brussels, trained as a physician. He set up a practice in Antwerp, where he wrote a book on the plague, before settling in Vilvoorde, near Brussels, with his family. His master treatise was *Ortus medicinae*, published in 1648, which considered the question of digestion and is sometimes regarded as anticipating the modern idea that food is broken down in chemical reactions conducted by enzymes.

See also: Experiment 14, The chemical composition of air and water, 1783 (page 66); Experiment 43, The chemistry of breathing, 1775–1790 (page 176).

by the German cardinal Nicholas of Cusa (1401–1464). Nicholas had suggested investigating plant growth by planting seeds in a pot filled with earth and watering them regularly. As the plant grew, said Nicholas, the mass of the earth would scarcely change—so the plants must get their weight from the added water.

There is no indication that Nicholas ever conducted the experiment. But van Helmont surely did. He took 200 pounds of earth, dried in a furnace and moistened with rainwater, and placed it in a pot in which he planted a willow sapling. He carefully covered the pot with a perforated iron lid to prevent dust from affecting his measurements. For five years he watered the plant and monitored its growth, after which he weighed it once more.

The soil had barely lost two ounces, but the sapling had gained 164 pounds, not including the weight of the leaves that had appeared and dropped with the seasons. All this mass, he said, "arose out of water only." Generalizing rather freely, he figured that this was true of all matter: "All earth, clay, and every body that may be touched, is truly and materially the offspring of water only."

This was clearly a long-term experiment, but van Helmont had time to spare. Because he had once published a treatise dealing with the healing power of religious relics that was deemed heretical by the Catholic church (which then held sway in the Low Countries), the Inquisition placed him under house arrest. This religious disapproval was also why van Helmont's willow-tree experiment was not published until after his death, appearing in his collected works *Ortus medicinae* (The Origin of Medicine) in 1648.

You can't fault van Helmont's method. Not content with just judging the growth by eye, he had carefully quantified his results: the hallmark of good experimental technique. Of course, his conclusions were quite wrong. Wood and bark, leaf and stem, never mind earth and clay and everything else, are not made only of water—although the greater part of the mass of all living things is comprised of the watery fluid that fills living cells. But van Helmont could hardly have supposed the truth of the matter: that plants get their substance out of the very *air*, by fixing invisible atmospheric carbon dioxide into sugars in the process of photosynthesis. It's a reminder that, even when the results of experiments are irrefutable, their interpretation may remain a tricky business.

Portrait by Mary Beale believed to be of Jan Baptista van Helmont, 1674, Natural History Museum, London.

The chemical composition of air and water (1783)

Q What are air and water made from?

The eighteenth century is often regarded as the age of "pneumatic chemistry": an era in which much of the focus of that discipline was on the "airs" or gases that could be produced in chemical processes. It was clear that these need not be the same as the common air that we breathe—and the question was how they differ. Thanks to an experimental method devised in 1727 by the English clergyman Stephen Hales, chemists could collect the "air" from a reaction by bubbling it through water and letting the bubbles accumulate in a submerged and inverted glass jar. Thus isolated, the properties of the gas could be studied.

In this way Joseph Black in Glasgow discovered what he called "fixed air," produced by heating limestone, which made lime water (a solution of calcium hydroxide) turn cloudy with a chalky precipitate. Black's student Daniel Rutherford showed that fixed air would not support life. In 1772, Rutherford also identified a different "malignant air" that remained after a substance was burned in ordinary air, and in which small creatures such as mice would also be asphyxiated. The Swedish chemist Carl Wilhelm Scheele found yet another "air" that bubbled off when a metal such as zinc was doused in acid, and which was inflammable, igniting with a pop.

How could all these different gases be explained? If you start, as eighteenth-century chemists would have done, with the assumption that ordinary air is itself a fundamental and irreducible element, you must conclude that the various chemical transformations are somehow changing it—adulterating it with other substances, say. Take combustion. It had been presumed since the start of the century that inflammable materials contain some substance that confers a propensity to burn, which is released into the air as combustion proceeds. In 1703, the German chemist Georg Stahl named this *phlogiston*. The idea was that air would accumulate phlogiston during combustion until it was saturated with it and could take no more, whereupon the burning would stop. This "fully phlogisticated air" was identified with Rutherford's "malignant air." Since it would not support life, this air was known in France as *azote*, which is Greek for "lifeless."

Gas balance used by Henry Cavendish for his experiments on the composition of air, ca. 1780, The Royal Institution, London.

William Alexander's graphite with gray wash *Sketch for a Portrait of Henry Cavendish*, late eighteenth century, The British Museum, London. The image was sketched somewhat surreptitiously, for the reclusive Cavendish had no desire for his portrait to be made.

Six months before Rutherford's account of malignant air was published, the English chemist and Nonconformist Joseph Priestley presented to the Royal Society in London an account of similar but more precise experiments conducted by one of its members, the wealthy gentleman Henry Cavendish. He was known to the Royal Society as an expert on pneumatic chemistry, but also as someone with whom it was very hard to engage on this or any other topic. One contemporary commented on how Cavendish would shuffle from room to room at the Royal Society, barely speaking to anyone and "seeming to be annoyed if looked at," but occasionally uttering a "shrill cry." He was the son of a lord and heir to a fortune, but always dressed in shabby, old-fashioned clothes.

Cavendish's experiments, which he conducted in his grand townhouse, near Piccadilly in London, were, however, models of precision. Not only did he record his measurements accurately, but he also understood the concept of errors, estimating the precision of his determinations and correcting for sources of error by, for example, taking the averages of many measurements.

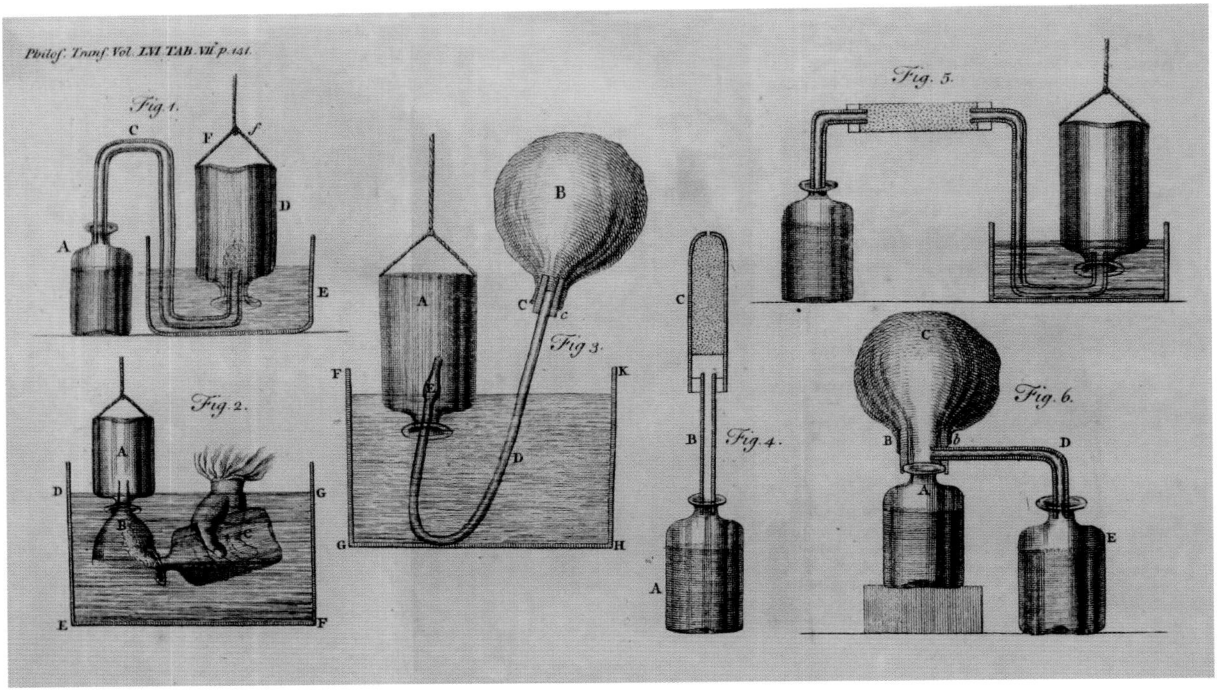

Factitious air apparatus. From Henry Cavendish's "Three Papers, Containing Experiments on Factitious Air," *Philosophical Transactions*, London: C. Davis, Printer to the Royal Society, 1766, Vol. LVI, Table VII, Natural History Museum Library, London.

Like Rutherford, Cavendish had looked at what becomes of common air in combustion. He passed ordinary air through a red-hot tube of charcoal, then removed any of Black's "fixed air" with lime water. What remained was an air with almost the same density as the original, but weighing 162 ounces to the original 180. To Cavendish, this gas was the residue of the "destruction of common air" and was fully phlogisticated by the combustion. But the French chemist Antoine Lavoisier later proposed that common air was, in fact, a mixture of two elements: oxygen (removed during burning) and the remaining *azote*. In 1790, his compatriot Jean-Antoine Chaptal proposed a new name for the latter element: *nitrogène*. As nitrogen, it caught on in England, but not in France.

Cavendish was fascinated by Scheele's inflammable air, which he weighed carefully and pronounced 8,700 times lighter than water. The English chemist John Warltire reported in 1774 that when this gas was ignited in air with an electrical spark in a sealed vessel, there was less "air" in the vessel afterward and dew coated the walls. Most chemists presumed that inflammable air was rich in phlogiston and that the explosive combustion and the "phlogistication" of common air made it lose the moisture it had previously contained. But Cavendish put numbers to this process: the amount of air remaining, he said, was four-fifths of the initial amount. As for the water that was formed, he had a complicated and rather ambiguous interpretation which supposed that the proportion of air lost in the process was, in fact, "dephlogisticated water." It was only once the phlogiston theory was dropped by Lavoisier and his followers that the proper interpretation became clear: "inflammable air" is a gaseous element that Lavoisier called hydrogen (the word means "water-former") and common air is a mixture of roughly four parts nitrogen to one part oxygen. When hydrogen is ignited in air, it combines with oxygen to form water, which is a compound of the two elements.

Cavendish's prowess in experimental science stemmed partly from his diligence and attention to detail, and partly from the fact that he could commission measuring devices from the finest instrument-makers in the land: balances, thermometers, barometers, hygrometers. Even the smallest anomaly did not escape his scrutiny. He noted that, while he could eliminate nitrogen (as we would now see it) from the air left after combustion by using sparks to react it with oxygen, there was always a tiny bit of "air" left that he couldn't remove: a recalcitrant little bubble. When his experiments were repeated in the 1890s, that residue was identified as a new, inert element: argon.

HENRY CAVENDISH | 1731–1810

Henry Cavendish was in some ways the archetypal gentleman-scientist of the late eighteenth century: born into an immensely wealthy, aristocratic family, with the means to pursue his own scientific interests at leisure. He was introduced to the Royal Society in the 1750s by his father Lord Charles Cavendish, who was also interested in the "experimental philosophy." But in other respects Cavendish was most unusual. He was probably on the autistic spectrum, and although he rarely missed a Royal Society meeting, he was usually too shy to speak and seemingly alarmed by social discourse. Yet he was a very careful experimentalist, and his studies made a vital contribution to our understanding of the physics and chemistry of "airs."

See also: Experiment 15, The discovery of oxygen, 1774–1780s (page 70); Experiment 43, The chemistry of breathing, 1775–1790 (page 176).

An eighteenth-century chemical laboratory, with furnaces, belonging to Ambrose Godfrey Hanckwitz (former laboratory assistant to Robert Boyle), ca. 1730s. Etching by William Henry Toms after Hubert-François Gravelot, Wellcome Collection, London.

The discovery of oxygen (1774-1780s)

Q How do things burn?

The phlogiston theory of combustion is often regarded as one of the most notorious detours in the history of science: a wrong idea that misled scientists for decades. But that betrays a lack of understanding of how ideas develop. Given the provisional nature of science, it is likely many of the things scientists believe today will eventually be replaced by something better. Yet ideas that do not stand the test of time may nevertheless act as bridges between ignorance and what we hope becomes, by stages, rather less of it.

Antoine Lavoisier was one of the many chemists for whom, in the early 1770s, phlogiston suppled a conceptual framework for rational experimentation. Like others, Lavoisier sought to understand the process called calcination, which happened when metals were heated in air. Typically, this made them lose their sheen and become a dull, brittle substance called a calx. The process could be reversed by heating a calx with charcoal.

The problem for phlogiston theory was that calxes typically weighed more than the metal itself. If the metal was giving out phlogiston as it was heated, why did it not become lighter? Some thought phlogiston might display negative weight, which was hardly a convincing explanation.

In around 1772, Lavoisier began to have other ideas. He proposed that metals "fix" some component of air when they become a calx. Hearing of Joseph Black's "fixed air" in 1773, he wondered if this was what combines with a metal to make a calx and was released when the calx is "reduced" by charcoal. But the French pharmacist Pierre Bayen pointed out that the "calx of mercury" could be reduced to mercury itself just by heating, and that this reaction released a gas that was not like fixed air. What then was it? In August 1774, the English chemist Joseph Priestley repeated the experiment and reported that the gas it generated made a candle flame burn brighter and turned a piece of smoldering charcoal incandescent. The next year, Priestley found that

Jacques-Louis David's oil on canvas *Antoine-Laurent Lavoisier and His Wife (Marie-Anne Pierette Paulze Lavoisier)*, 1788, The Metropolitan Museum of Art, New York.

this gas could keep mice alive in a sealed jar for longer than could ordinary air. When he inhaled it himself, Priestley reported that "my breath felt peculiarly light and easy for some time afterward."

Priestley was a believer in phlogiston theory, and he regarded this new "air" as "dephlogisticated air": because all phlogiston was removed from it, it could absorb it more readily from burning substances. Unbeknown to him at the time, the Swedish apothecary Carl Wilhelm Scheele had produced the same gas by different chemical means in 1771–1772 and called it "fire air."

In October 1774, Priestley dined with Lavoisier in Paris and mentioned his findings. This helped persuade Lavoisier that calxes were not a combination of metals with fixed air after all, but with another gaseous substance. By 1775 he had carried out the same experiments with calx of mercury. At first he thought the heated calx was releasing ordinary air, but after Priestley sent Lavoisier a sample, the French chemist decided it was instead an especially "pure air." Gradually, he concluded that it was, in fact, a new chemical element in its own right, and that common air was a mixture of this and another substance.

Lavoisier's key experiment to verify this hypothesis, presented to the French Academy of Sciences in 1777, involved cycling the "more respirable" air into and back out of a calx. He heated mercury in air until it would calcine no more, then used a "burning lens" to heat the calx and release the new air, which he recombined with the unreactive residue (as we'd now see it, the nitrogen) left from the original calcination, showing that this regenerated common air "pretty exactly." The force of reasoning rested on the demonstration that nothing was lost, which demanded considerable experimental skill.

Having combined this new gas with other elements, such as sulfur, carbon, and phosphorus, Lavoisier decided it was the "principle of acidity": the element that gave rise to acids (like sulfuric). This supposition was wrong—not all acids contain the new element—but it motivated him to name the element *oxygène*, or "acid-maker." Given oxygen, phlogiston was not needed at all. When substances burn, Lavoisier said, they do not release phlogiston, but rather take up oxygen from the air. Abandoning phlogiston altogether was too great a step for many chemists, including Priestley. But while it was received with scepticism in England for the rest of the eighteenth century, the theory found supporters in France. Lavoisier and his colleagues effectively staged a coup by developing a new system of chemical nomenclature that implicitly encoded the oxygen theory: calxes, for example, became oxides. No single experiment proved the theory, and, indeed, some of the crucial experiments, like the production of oxygen from calx of mercury (mercuric oxide), were not his at all. But Lavoisier couldn't have made a compelling case for his interpretation had he not been so meticulous, weighing reagents and products carefully to measure what had been lost or gained.

ANTOINE LAVOISIER
1743–1794

Born in Paris to a wealthy family, Antoine Lavoisier became a tax administrator under Louis XIV and was appointed a "Fermier General" in 1780. He pursued his chemical studies at the same time, setting up a laboratory in his home. In 1775 his skills earned him a place on the Royal Gunpowder Commission. Although Lavoisier began serving the Republic after the French Revolution of 1789, his former job as a tax collector left him vulnerable to accusations of collusion with the old regime. He was arrested and imprisoned in 1793 on charges of defrauding the state and sent to the guillotine the following year. The famous response of the court to the appeal for clemency made during his trial—"The state has no need of savants"—is almost certainly apocryphal, but it surely captures the mindless destructiveness into which the revolution descended. Within a decade of his death, Lavoisier's new vision of chemistry had transformed the discipline.

The discovery of alkali metals by electrolysis (1807)

Q Can electricity be used to separate compounds into their elements?

In 1661, the Anglo-Irish scientist Robert Boyle declared that a chemical element is any substance that can't be broken down into simpler ones. Such elements are the fundamental ingredients of the physical world.

The problem is that it can be hard to know if a candidate substance really is irreducible, or if you just haven't found the right way to split it into its components. In a list of elements published in 1789, French chemist Antoine Lavoisier included soda, potash, "lime" (chalk), magnesia, alumina, silica, and others: minerals that no one had ever succeeded in breaking down into anything more fundamental.

However, in the early nineteenth century, the young chemist Humphry Davy suspected that these substances contained hitherto unknown metallic elements, which were presumably bound very tightly to others and not easily parted. Davy surmised that they might be prised apart with a dramatic new method: passing an electrical current through the compounds.

Experimenters had only recently begun to tame electricity. From the mid-eighteenth century they learned how to generate it as static

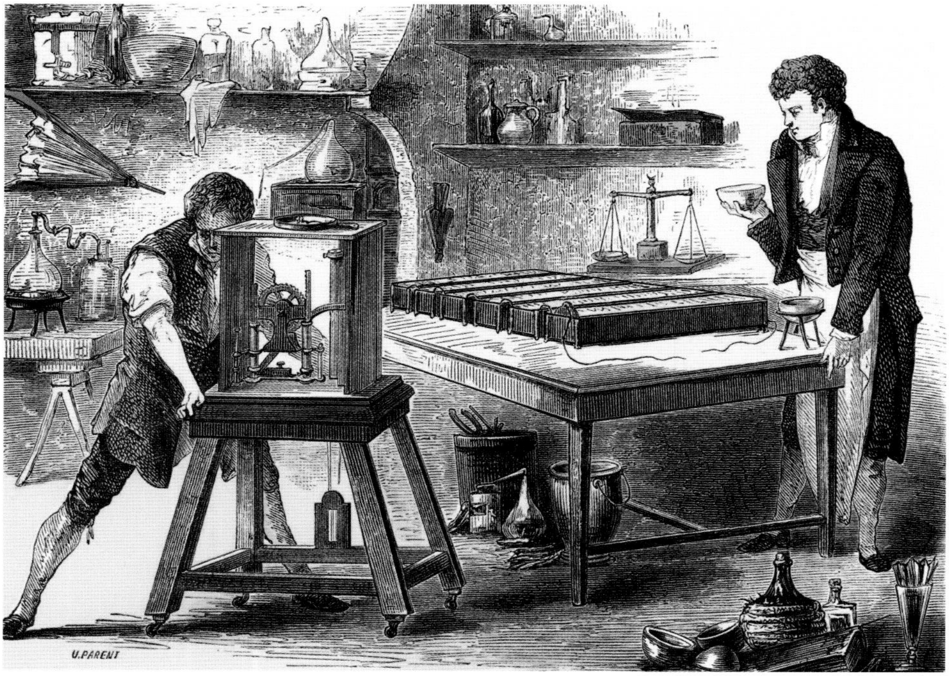

Nineteenth-century color engraving showing Humphry Davy using electrolysis to discover potassium and sodium.

Apparatus used by Humphry Davy to isolate the metal potassium from "caustic potash," 1807, The Royal Institution, London.

electricity—for example, by rubbing a glass sphere—and to store it in "bottles" called Leyden jars. In 1800, the Italian scientist Alessandro Volta, in Pavia, found that he could generate an electrical current by stacking disks of two different metals, such as copper and zinc, in an alternating sequence and separating each disk from the next with salt-soaked cloth or card. This structure, which became known as a voltaic pile, was, in fact, the first battery.

In the same year, the English scientists William Nicholson and Anthony Carlisle found that when they passed the current from a voltaic pile through water, they could split the liquid into its constituent elements, hydrogen and oxygen, which bubbled off as gases at two metal electrodes inserted into the liquid to conduct the current. They had split water using electricity, a process that became known as electrolysis.

Might this same technique split substances like potash and soda into hitherto unknown elements, Davy wondered? As an assistant to the Bristol-based physician Thomas Beddoes, this ambitious young man repeated some of Volta's own experiments with the pile, and in 1801 he was appointed as laboratory director at the Royal Institution in London, established as a center for experimental science in 1799 by Sir Joseph Banks. There Davy was given a pile for his studies

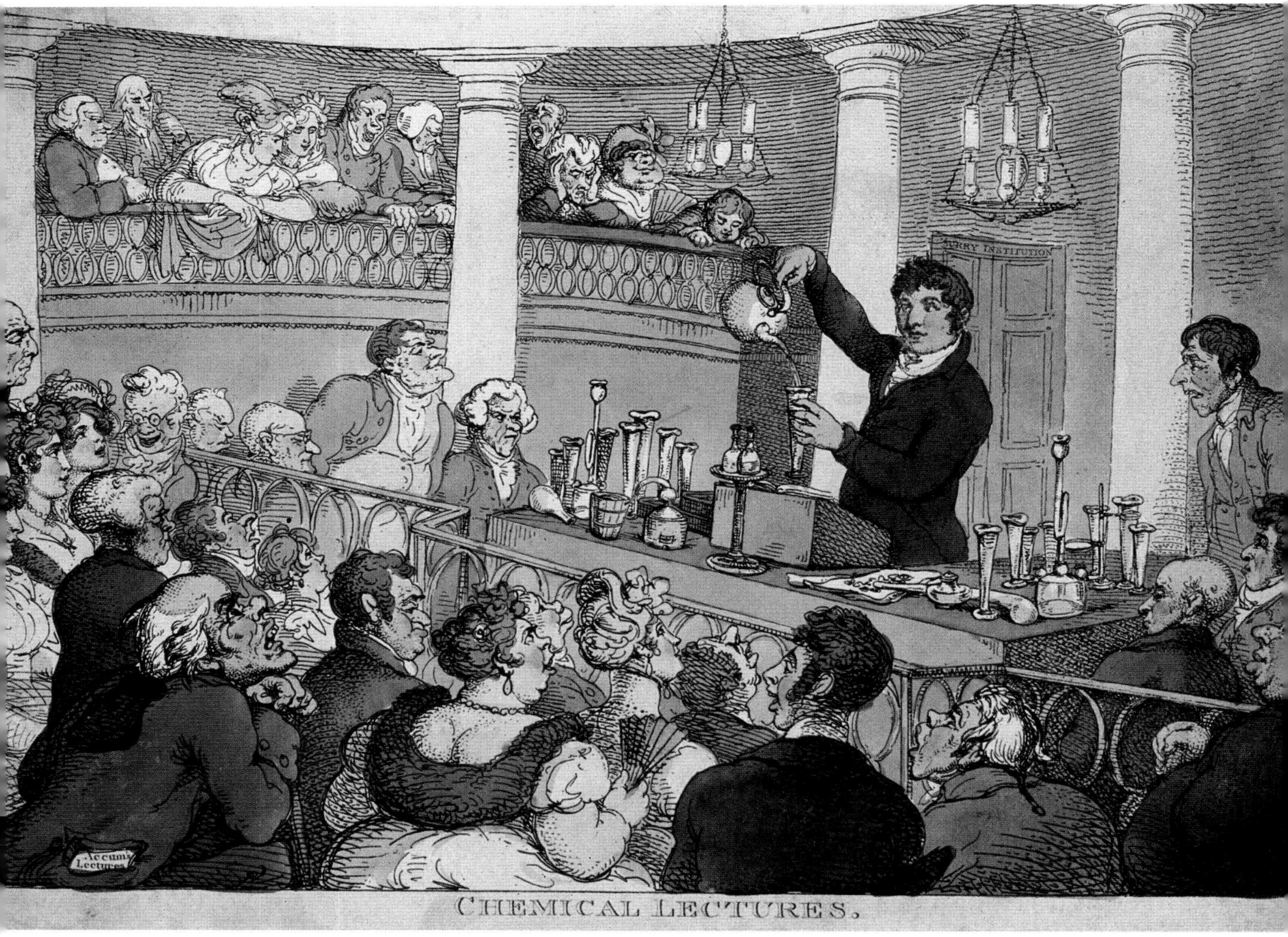

Thomas Rowlandson's colored etching of a chemical lecture at the Surrey Institution, 1809, Wellcome Collection, London. Davy is the man on the far right between the columns.

by Volta himself; it was at that stage the most powerful battery in the world.

Davy found that some metals, such as iron, tin, and zinc, could be extracted from solutions of their salts by electrolysis. The metals would coat the surface of the electrode connected to (as we would now see it) the negative terminal of the primitive battery. However, no metal was forthcoming from a solution of potash (potassium hydroxide). Instead, hydrogen bubbled off from the negative electrode.

Maybe Davy needed to get rid of the water from which the hydrogen came? In 1807, he tried instead to electrolyze pure potash, heated until it melted. This time he saw small, metallic-looking globules form at the negative electrode (made of platinum). This substance had dramatic properties. Oddly for a metal, Davy reported, "it is very soft, and easily moulded by the fingers." Sometimes it would combust spontaneously with a bright flame, and it would ignite "with great violence" when put into water, burning with a flame of "white mixed with red and violet" while moving about on the surface of the water.

Davy proposed that this substance was a new metallic element, for which he suggested the name potassium. It is a highly reactive metal, which combines explosively with water to produce flammable hydrogen gas, as the potassium atoms shed electrons to become positively charged ions. That is why it won't form by electrolysis of potash solution: it is just too reactive. At the negative electrode, some element must take up electrons delivered by the electrical current—but the potassium ions in the solution are too resistant to this, and so hydrogen atoms plucked from water molecules must do so instead, forming hydrogen gas.

Davy naturally turned his attention next to soda, in the form of "caustic soda" (sodium hydroxide), known also as lye. Again, electrolysis of the molten salt produced a metal at the negative electrode—"as white as silver," Davy wrote, and reactive like potassium, but slightly less so: it is less apt to burst into flame when added to water. He named it sodium. In 1808, Davy experimented with the electrolysis of lime, magnesia, and another of Lavoisier's "earth" elements, called baryte, and in this way he discovered three new metals: calcium, magnesium, and barium, as well as a fourth—strontium—extracted from the mineral strontia, which had been identified in 1790 in a Scottish lead mine. Thus, Davy's electrolysis experiments revealed, within the space of just a few years, a cluster of new metals, along with the elements boron, aluminum, silicon, and zirconium. He was not, however, always able to produce these in pure form; other researchers did so later in the nineteenth century.

HUMPHRY DAVY | 1778–1829

Born in Penzance, Cornwall, Humphry Davy began experimenting in chemistry as the apprentice to a local apothecary. He met his mentor Thomas Beddoes when Beddoes came to Cornwall on a geological expedition, and subsequently became his assistant in Bristol, before being appointed assistant lecturer in chemistry at The Royal Institution in London. There he became an expert in "galvanism," the science of electricity, as well as devising the first incandescent lamp and the miner's safety lamp that bears his name. When Davy was elected president of the Royal Society in 1820, he was arguably the most celebrated scientist in England.

See also: Experiment 15, The discovery of oxygen, 1774–1780s (page 70); Experiment 19, The discovery of radium and polonium, 1898–1901 (page 88).

With this set of experiments, Davy uncovered a truth more profound than the existence of new elements: that chemistry was fundamentally electrical in nature. That's to say, chemical reactions involve the redistribution among the component atoms of electrons, the negatively charged constituents of atoms whose motion constitutes an electrical current. By harnessing electricity to control these transactions in electrons, Davy launched the discipline of electrochemistry. But the principle is deeper still, for in fact the chemical bonds that unite all atoms in molecules and compounds are forged and broken as electrons are transferred between them.

The "handedness" of molecules (1848)

Q Why do some crystals have a handedness to their shape?

"Fortune," the French chemist Louis Pasteur famously said, "favours the prepared mind." Plenty of important discoveries in science have come about through sheer serendipity, being revealed via some phenomenon or result that was never sought at the outset and which might even have been elicited by an error in the experimental process. But what turns such an outcome into a discovery, rather than an experiment that is simply considered to have failed and which is discarded, is a mind honed by experience and the training to see the result as worth pursuing and as potentially disclosing something new. The best scientists have a sense of when to pay attention to an anomaly or an unexpected result, and when to ignore it.

Ironically, we should take Pasteur's dictum with a grain of salt when it is applied to himself. He was undoubtedly one of the finest scientific minds of his generation in France, but he was also apt to retell the history of his work to reflect well on his own astuteness. That seems to be the case for his most celebrated experiment, in which Pasteur discovered that molecules themselves can have a "handedness": they may have alternative, mirror-image spatial arrangements of atoms. Pasteur's experiment itself has considerable elegance; it's just that we can't be sure that he set out to discover quite what he did.

At the heart of the matter was the way light interacts with matter. In the early nineteenth century, French scientists discovered that light can be polarized: as we'd see it now, the oscillations of the light waves are all in the same plane. In 1815, Jean-Baptiste Biot found that when polarized light passes through some crystals, the plane of polarization may be rotated by a fixed angle. What's more, this so-called "optical activity" may remain when the crystals are melted or dissolved—so the behavior can't stem from the way the molecules are arranged in the crystals but must be intrinsic to the molecules themselves.

Pasteur studied this phenomenon for his doctoral thesis at the *École Normale Supérieure* in Paris. What did it imply for the shape of the molecules? In 1848, Pasteur came across a paper by Biot reporting two types of tartaric acid, an organic (carbon-based) acid produced during wine-making. Normal tartaric acid was optically active, but a chemically identical variant called

LOUIS PASTEUR | 1822–1895

Louis Pasteur has been called one of the first microbiologists and the founder of modern bacteriology. Together with the German physician Robert Koch, he established the germ theory of disease: the notion that diseases may be caused by invisibly small microorganisms called bacteria. Best known publicly for his invention of the process dubbed pasteurization for treating milk against bacterial contamination, he also performed experiments that discredited the idea of spontaneous generation of living matter (see page 190) and undertook pioneering work on vaccination.

See also: Experiment 18, The three-dimensional shape of sugar molecules, 1891 (page 84).

Photograph of Antoine Balard's laboratory at the *École Normale Supérieure* where Louis Pasteur worked from 1846 to 1848 and made his discovery of molecular asymmetry in 1848, Institut Pasteur, Paris.

racemic acid was not, Biot said. He claimed that crystals of these compounds were identical, but Pasteur synthesized both types of crystal in the laboratory and looked at them closely under the microscope. He saw that the faceted crystals are not symmetrical, but have a handedness, like seashells that twist to the right or left. While crystals of salts of tartaric acid always had a right-handed twist, those of racemic acid could be either left- or right-handed.

Seeing this, Pasteur later claimed that "for an instant my heart stopped beating." Rather than being optically inactive—not rotating polarized light at all—might racemic acid in fact be an equal mixture of left- and right-handed tartaric acid molecules whose effects cancel out, but that separate when they crystallize and produce crystals of opposite handedness?

Pasteur could find out. Using tweezers, he painstakingly sorted the racemic salt crystals

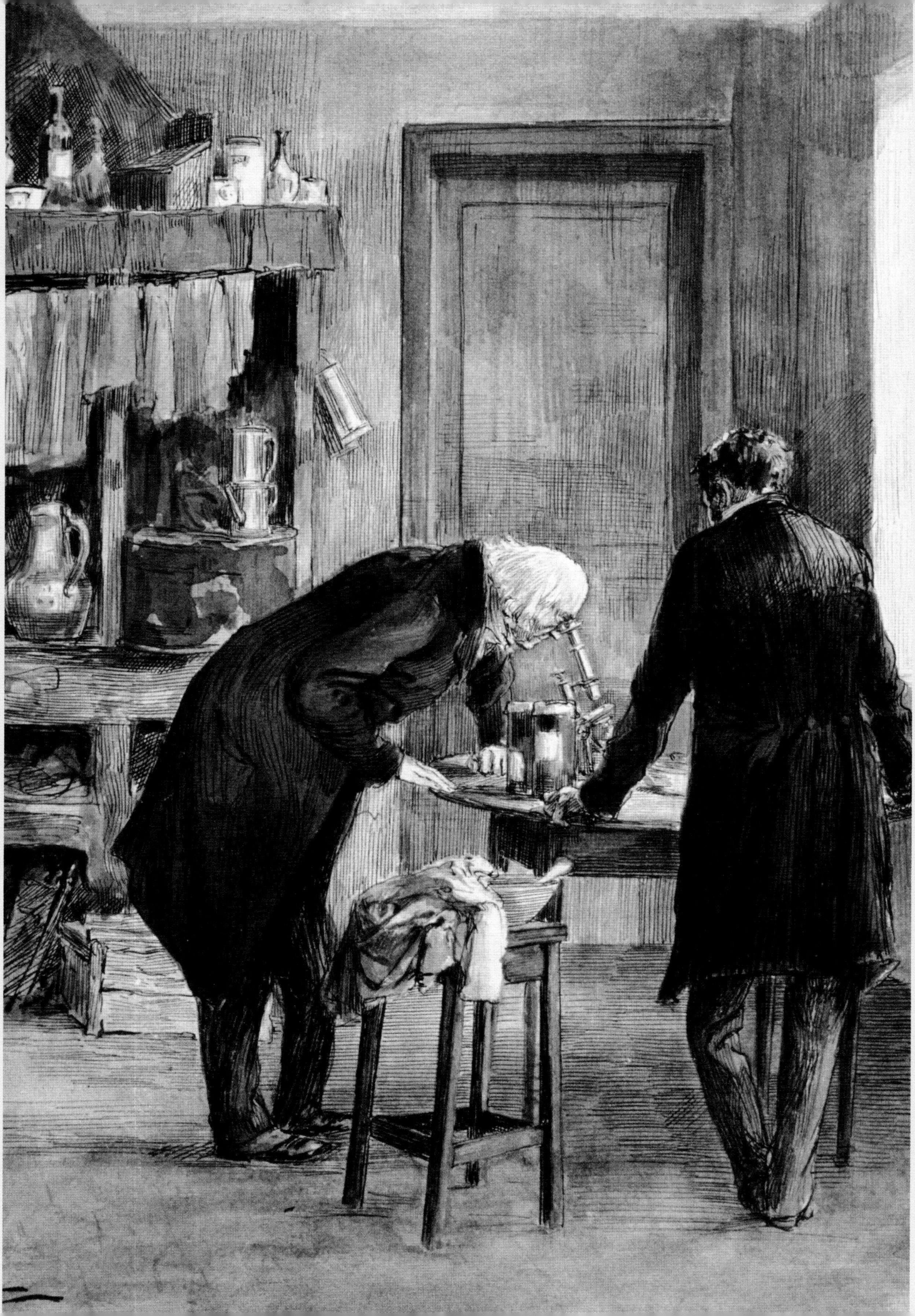

Opposite: Hermann Vogel's illustration of Jean-Baptiste Biot observing Louis Pasteur's studies on crystallography through the microscope in his laboratory at the Collège de France in 1848, Institut Pasteur, Paris.

into two piles: left- and right-handed. Then he redissolved the piles separately and found that the resulting solutions did indeed rotate polarized light in opposite ways. Later accounts of the experiment had Pasteur exclaiming: "All is found!"—a Gallic "Eureka!"—and rushing out to tell the first person he encountered. There's probably some romantic mythologizing here, not least because Pasteur's notebooks show that his initial findings were not entirely clear-cut.

All the same, the reasoning was sound and the discovery significant. The implication was that molecules of tartaric acid do indeed have mirror-image forms: a kind of left- and right-handed twist. How can that be? Perhaps, Pasteur wondered, the constituent atoms are arranged in a corkscrew-like helix, which can twist like the thread of a screw to the left or right. Another possibility, he suggested in 1860, is that the atoms sit at the vertices of a tetrahedron: for four different types of atom, there are two mirror-image ways of making that arrangement.

It was a good guess. In 1874, the Dutch scientist Jacobus van't Hoff proposed that carbon-based molecules may indeed have that structure: each carbon atom may sit at the center of a tetrahedron and form chemical bonds with four other atoms or groups of atoms disported in space around it. In 1904, Lord Kelvin gave such handed molecules a name: chiral, from the Greek *kheir*, meaning "hand." Some key biological molecules are chiral, such as the amino acids that make up proteins and the (right-spiraling) double helix of DNA.

Pasteur's good fortune was that the two chiral forms of tartaric acid will spontaneously separate when the compound crystallizes, forming the handed crystal shapes, because the molecules can pack more efficiently that way in the crystal lattice. Most chiral compounds won't separate so readily, however: their crystals are mixtures of both molecular forms. What's more, this separation only happens for racemic acid when it crystallizes in cold conditions. Pasteur did his experiments in the winter in an unheated lab. Had he done so during the Parisian summer, he might not have been so lucky.

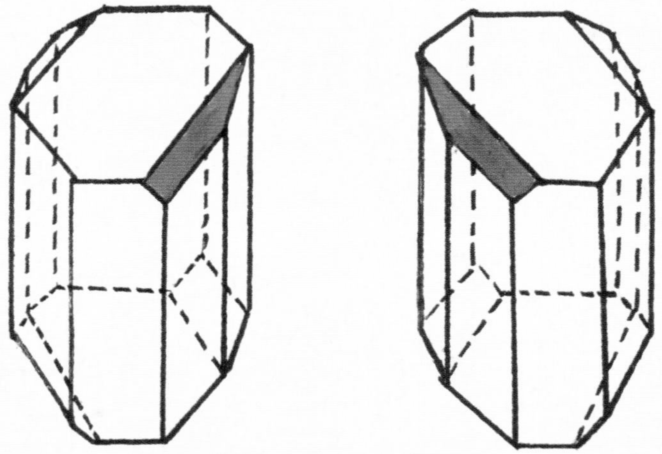

The right and left crystals of sodium ammonium paratartrate, showing chirality, after the original drawing by Louis Pasteur, early twentieth century, Institut Pasteur, Paris.

What is a beautiful experiment?

It might surprise many people to discover that "beauty" is a word more likely to be spoken today by scientists than by artists. There is almost a sense among contemporary artists and art theorists that the word is unseemly, perhaps even untrustworthy. Scientists, meanwhile, wax lyrical about "beautiful theories"—and beautiful experiments, too. Many claim that this aesthetic reaction is no different to that elicited by art, but it is hard to pin down exactly what it consists of or how it is evoked. Some scientists associate beauty with symmetry—a feature central to modern physics—but they would struggle in vain to reconcile this idea with aesthetic theories in art. It is challenged, for example, by Immanuel Kant's claim that "All stiff regularity (such as approximates to mathematical regularity) has something in it repugnant to taste": we quickly weary of its simplicity. And while some scientists assert that their notion of beauty is timeless and universal, few would claim the same for art.

Some philosophers have argued that "beauty" in science merely stands as a proxy for truth: what is true is then necessarily beautiful. If that is so, such allegedly aesthetic judgments seem a little shallow, and also perilous: we might be tempted to place undue trust in an idea simply because we deem it beautiful. Some scientists have, however, defended that position. The British physicist Paul Dirac (see page 126), for example, claimed that it is more important that a theory be beautiful than that it conform with experiment, while Einstein stated that "the only physical theories that we are willing to accept are the beautiful ones." Others are sceptical that perceptions of beauty are any guide to validity: the zoologist Thomas Henry Huxley said that the "great tragedy of science" is "the slaying of a beautiful hypothesis by an ugly fact."

One argument for why beauty is a valid descriptor in supposedly objective science is that the aesthetic response in scientists seems to involve the same neural pathways as those stimulated by reactions to art. But this doesn't prove as much as it seems. After all, the same reward circuits of the brain are activated by sex, food, and music—but this hardly implies that they are all essentially the same activity or that one may substitute for the other. There is no evidence that our brains possess some kind of innate neural "beauty circuit."

In contrast to notions of "beautiful theories" (an assessment commonly applied, for example, to Einstein's theory of general relativity), it's less obvious that beauty pronounces at all on the outcome of an experiment. Rather, that tends to be a judgment applied before the outcome is known, and is made more on the grounds of the design and logic embodied in the procedure. As French physicist and philosopher of science Pierre Duhem said, experiments may be seen as embodied hypotheses—and there's an attraction to an experiment that performs the translation efficiently and unambiguously, as, for example, in Ernest Rutherford's study of the alpha particle (see page 108). All the same, there's likely to be some post-hoc justification in those experiments commonly designated as beautiful. There are probably many beautifully planned and executed experiments that have been forgotten because they didn't work, or not well enough, or not in a way that could be easily interpreted.

Frontispiece from Francis Bacon's *De augmentis scientiarum*, Leiden: Ex Officina Adriani Wijngaerden, 1652, Book IX.

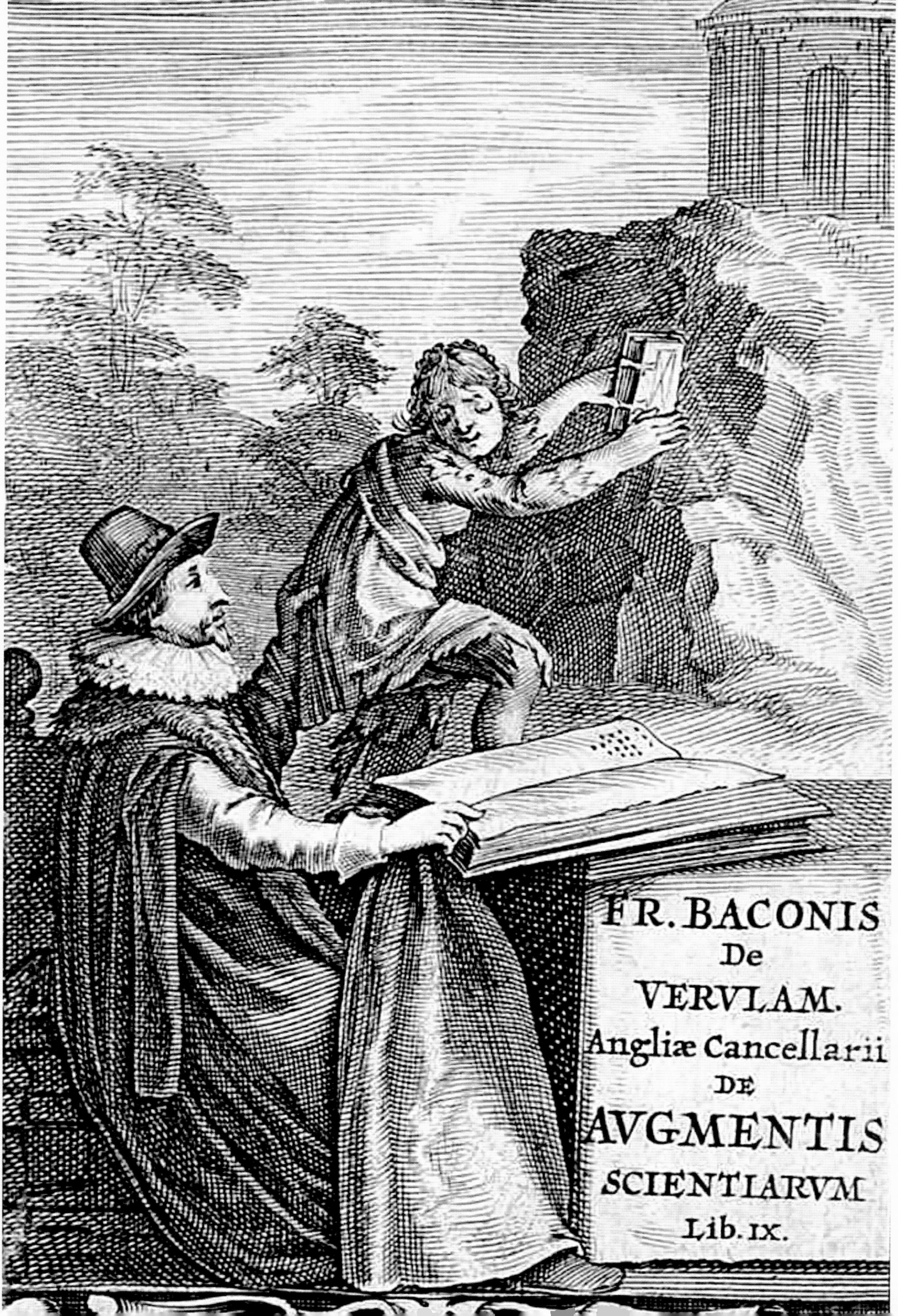
FR. BACONIS
De
VERVLAM.
Angliæ Cancellarii
DE
AVGMENTIS
SCIENTIARVM
Lib. IX.

Facing facts: Thomas Henry Huxley lecturing on the skull of the gorilla, using a pullout blackboard, ca. 1861, photographed by Cundall Downes & Co., Wellcome Collection, London.

With that proviso, the beauty awarded to an experiment lies more in its execution than in its outcome. It has something in common with the (expert) appreciation of a game of chess, arising from the aptness of the moves, the elegance of the strategy, and the choices that force the opponent's hand. While Francis Bacon in the seventeenth century notoriously implied that experiments subject nature to a degree of coercion, an elegant experiment can look more like a collaboration of the experimenter with nature to uncover "something deeply hidden," as Einstein put it. A beautiful experiment marshals the available resources to disclose what casual inspection will not. What leads many biologists to consider the Meselsohn–Stahl experiment (see page 200), which revealed the replication mechanism of DNA, the most beautiful in their discipline was how a seemingly impossible puzzle—to distinguish between possibilities whose outcomes look identical—was turned into a soluble one.

Beauty as a means of learning and exploring

In experiments there are many potential ingredients of aesthetics: beauty of concept, beauty (especially economy) of instrumental design, the aptness and economy with which the two are aligned, and beauty of reasoning in interpreting the results.

These are qualities that require creativity and imagination—there is no prescription for them. Some scientists seem to have a talent for aesthetically pleasing experimental design, and none more so than Rutherford. Such virtues are perhaps easier to spot in an experiment than in a theory, for they don't tend to require recondite knowledge and are, as it were, explicitly built in.

Physics Nobel laureate Frank Wilczek, author of the 2015 book *A Beautiful Question: Finding Nature's Deep Design*, has suggested that beauty in a scientific idea becomes manifest when "you get out more than you put in": the idea delivers something new and unexpected, revealing more than anticipated. It's an intriguing thought when applied to experiments, for in comparison to theories the "deliverables" of experiments are more explicit: often a simple yes/no or this/that answer. Yet one can find examples of such bountiful excess in an experiment's answers. Consider, for instance, the crystallographic studies that guided James Watson and Francis Crick to solving the molecular structure of the DNA molecule in 1953. That double-helical structure is widely considered beautiful in its own right—both Crick and Watson used the word, although convention forbade it in print—but it also, as the pair famously mentioned rather archly in their discovery paper, showed how DNA might be replicated when cells divide (see page 204). No one expected the structure to so obviously present a solution to that question too.

Perhaps we shouldn't try too hard to pin down notions of beauty in science: attempts to make it a parameter that we can quantify and measure are liable to kill it as surely as vivisection kills the unfortunate lab animal. At any rate, the beautiful experiment, like the beautiful theory, probably gains in persuasive power: why, of course nature is like that! There's a danger there—we shouldn't be blinded by beauty. But beautiful experiments tend almost by definition to be *good* experiments: they have clarity, they are unambiguous, and they deploy the available means in a logical and well-ordered fashion. This is surely how experimenters should aspire to work: beauty here serves a pedagogical function too. Not all of the experiments described in this book are "correct" in modern terms, but this does not detract from their aesthetic value. All science must be of its time, and good science can and does produce answers that are later revised and replaced.

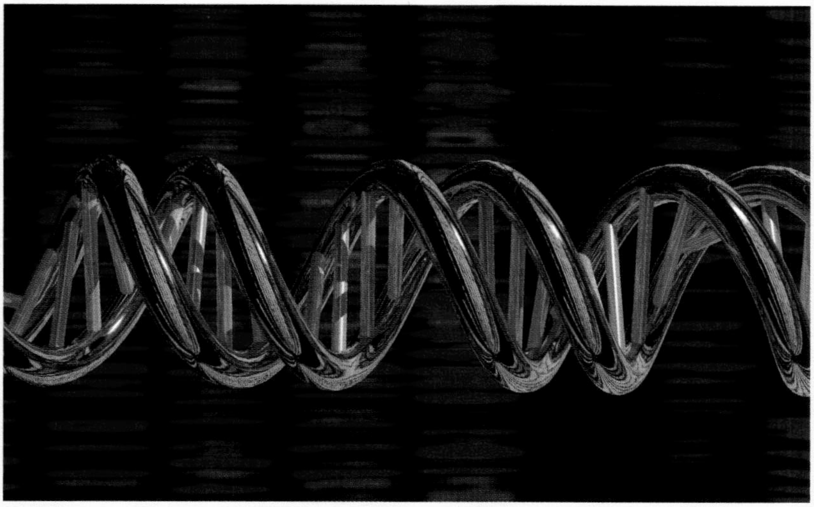

Artistic representation of the double helix of DNA, displayed against a background of genome-sequencing data, Peter Artymiuk, Wellcome Collection, London.

The three-dimensional shape of sugar molecules (1891)

Q What distinguishes one type of sugar from another?

The chemistry of carbon compounds is an entire subdiscipline called organic chemistry because many carbon compounds are found in living organisms. All life on Earth is typically described as "carbon-based," and it is the ability of carbon to form the basis of such a diverse array of molecules that makes this life possible.

But that very diversity made organic chemistry deeply confusing to chemists in the nineteenth century, when they became adept at quantifying which elements substances contained and in what proportions. What were the rules governing the nature of carbon compounds? How does one carbon-based molecule differ from another? Chemists identified distinct families of carbon compounds with similar properties: those called alkanes (containing carbon and hydrogen and found in crude oil), say, or carboxylic acids, proteins, "aromatic" compounds such as benzene (often found in coal tar), and sugars.

From around the mid-nineteenth century, researchers recognized that molecules consist of atoms linked together into different structures and shapes. The identity and the properties of a given compound are defined by how its atoms are joined and arranged in space. It is possible not just for the same elements to be linked in different ways and proportions, as in the alkanes, but also for molecules with precisely the same numbers of atoms of each element to have different structures: different arrangements of those atoms in space. The latter are called isomers.

In 1874, the Dutch chemist Jacobus van't Hoff discerned the key structural motif of carbon compounds: in general, he said, carbon atoms prefer to form chemical bonds to four other atoms, arranged such that the carbon sits at the center of a tetrahedron with its four neighbors at the vertices. If the four atoms attached to the carbon are different from one another, van't Hoff (and, independently, Joseph Le Bel in France) pointed out, there are two distinct ways of arranging them around the vertices, one of which is a mirror image of the other. Such alternatives are called enantiomers. The carbon atom at the center of these arrangements is said to be chiral, which means "handed" (see page 76). For molecules with several chiral carbon atoms there may be several different structural arrangements, called stereoisomers.

Armed with this knowledge, the German chemist Emil Fischer set out to understand what it is that distinguishes one sugar—a "carbohydrate" compound, containing just carbon (C), hydrogen (H), and oxygen (O) atoms—from another.

Several sugars were then known, often named from the substances they were extracted from: fructose (in fruits), maltose (in malt), lactose (in milk), and so on. They all seemed to have similar properties—all, for instance, tasted sweet. This was, in the words of historian of science Catherine Jackson, "an extreme problem of chemical identity"—it was very hard to figure out whether two substances were the same or different, and if the latter, in what way.

Fischer learned his craft in the 1870s in the laboratory of Adolf von Baeyer, one of the most accomplished organic chemists of his times, at the University of Strasbourg, before moving to

Emil Fischer at a laboratory table, 1904, National Library of Medicine, Bethesda, Maryland.

EMIL FISCHER | 1852–1919

The German Hermann Emil Fischer is regarded as one of the founders of modern organic chemistry. At the University of Strasbourg in the early 1870s he studied under another giant of the field, Adolf von Baeyer. He worked on synthetic organic dyes (then a huge industrial concern), discovered one of the first barbiturate sedatives, barbital, and made pioneering contributions to the understanding of the chemistry of proteins. He was awarded the second Nobel prize in chemistry, in 1902, for his work on sugars. (The first went to the Dutch chemist Jacobus van't Hoff.)

See also: Experiment 17, The "handedness" of molecules, 1848 (page 76).

the University of Munich, then Erlangen in 1881 and Würzburg in 1885. He became an expert in separating and characterizing natural organic molecules such as caffeine, proteins, and the substances excreted by animals, and also in making (synthesizing) such compounds by reacting together simpler chemical ingredients.

Fischer began his studies of sugars in 1884. These molecules, we now know, share a common motif: a ring of five or six atoms, all of them carbon except for one oxygen. Attached to the edges of the ring are various hydroxyl (O-H) groups. In monosaccharide sugars like glucose and mannose there is just one ring, but disaccharides like sucrose (common culinary sugar) have two. In starch, many sugar rings are linked in a chain called a polysaccharide. Confusingly, when sugars are dissolved in water their rings can spring open by a rearrangement of the chemical bonds, producing an "open-chain" isomer called an aldose. The aldose structure was already known before Fischer's work; Baeyer himself had suggested as much.

Many of the carbon atoms in both the cyclic and the open-chain isomers of sugars are chiral, creating a variety of possible stereoisomers. That's a big part of what makes it so challenging to figure out the structures of the sugars.

Fischer's strategy was to see if he could synthesize the different sugars from simple starting materials. If there is only one way

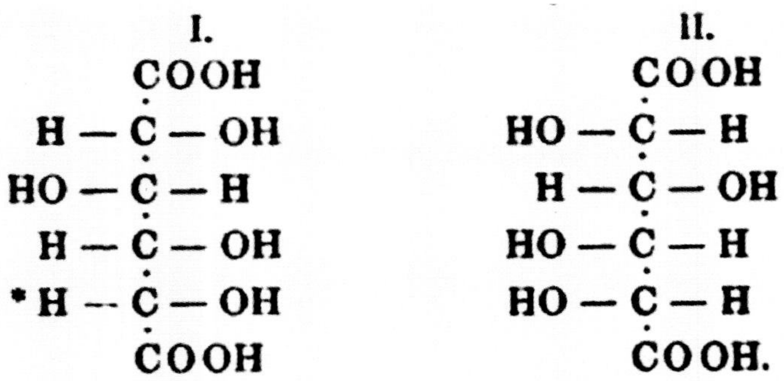

Configuration of sugars, from E. Fischer. 'Über die Configuration des Traubenzuckers unde Seiner Isomeren (On the configuration of glucose and its isomers)', Berichte der Deutschen chemischen Gesellschaft zu Berlin, 1891, vol. 24 (2), p.2684. University of California.

Students thought to be working on the configurations of sugars at the Chemical Institute lecture hall, Friedrich-Wilhelms-Universität zu Berlin (Humboldt University), ca. 1904.

those initial compounds can link up, it's possible to deduce what structure they must produce. In order to deduce the identity of the product, Fischer had only relatively crude measures: in particular, he could make crystals and then measure their melting point and compare it to the melting points of known sugars (which are different for different sugars). Alternatively, Fischer would cause the sugars themselves to react with other molecules or to fall apart, and work out from the nature of the products what the original configuration of atoms must have been. In these chemical manipulations Fischer was particularly reliant on a compound he discovered in 1875, called phenylhydrazine, which he reacted with sugars to make products that could be crystallized and compared.

By 1888 Fischer had worked out the structures of the closely related sugars glucose, mannose, and fructose, and their stereochemical structures. Two years later, he had figured out how to make the three compounds from glycerol. And by 1894 he had deduced the structures and the relationships between all the known sugars.

This was, then, not a single experiment but an entire experimental program that extended over about a decade, based on a common set of methods. It established one of the organic chemist's most valuable tools over the century that followed: working out the structure of a complex molecule by synthesizing it in a series of small steps, the outcome of each of which could be predicted. Jackson calls Fischer's work on sugars a classic example of "laboratory reasoning": the process by which chemists translate the chemistry they carry out in flasks in the lab into an understanding of the constitution and structure of molecules. Crucial to Fischer's success was his experimental prowess at creating molecules, coupled with his ability to decipher what the experimental results implied for the molecules whose forms and shapes were beyond the means of any microscope to discern. What the chemist worked with were solutions, crystals, beakers, wet stuff; what they visualized in their predictions and interpretations were atoms and molecules.

The discovery of radium and polonium (1898-1901)

Q Is there a new radioactive element in uranium ore?

Radioactivity, discovered at the end of the nineteenth century, defied conventional understanding. These energetic rays emanating from certain substances seemed to be inexhaustible, and yet there was no known source of energy that might produce them. Even more heretically, the phenomenon appeared to result in the transmutation of one element to another: an unheard-of process, with uneasy echoes of long-discredited medieval alchemy.

The discovery came about by chance. It was stimulated by another shocking revelation: the existence of X-rays, discovered by the German scientist Wilhelm Röntgen in 1895. Röntgen detected these invisible rays while studying so-called cathode rays emitted by negatively charged metal electrodes. He observed that a phosphorescent screen that happened to be placed near a cathode-ray tube would glow, and that the rays apparently emanating from the tube to cause the glow could pass through paper and even through flesh and blood. Hearing of the discovery in 1896, the French physicist Henri Becquerel decided to investigate whether phosphorescent materials, which glowed when placed in sunlight, also emitted X-rays. One such was a salt of the metal uranium, an element discovered in 1789.

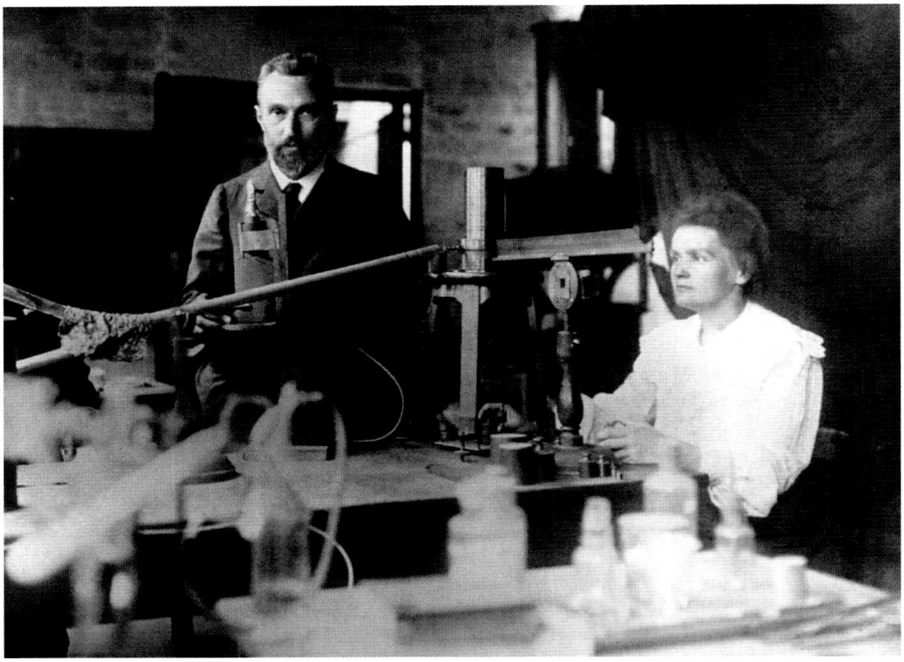

Pierre and Marie Curie photographed in their Paris laboratory, ca. 1898, Musée Curie, Paris.

X-rays could be revealed by the way they would darken a photographic emulsion, and Becquerel found that a photographic plate wrapped in black paper would indeed be darkened when the salt was placed on top and exposed to sunlight.

One cloudy day, Becquerel placed one such plate in a cupboard with the salt on top, intending to return to it when the weather improved and it could be exposed to sunlight. After more days of poor weather, he removed that plate and, seemingly on a whim, developed it anyway—to find that the uranium salt had darkened it despite having seen no bright sunlight. Evidently, the effect did not depend on exciting phosphorescence: uranium spontaneously emitted rays of its own accord. Unlike cathode rays, X-rays, or phosphorescence, these "uranic rays" did not need a source of energy to stimulate them.

It was deemed a curiosity and little more, but when Marie Curie, a Polish woman studying at the École Supérieure de Physique et Chimie Industrielle in Paris, needed a doctorate topic in 1898, uranic rays were appealing: "the question was entirely new," she wrote, "and nothing yet had been written upon it."

Born Maria Skłodowska, she had married the French scientist Pierre Curie in 1895, four years after arriving in Paris. The Curies approached the study of uranic rays together: Pierre devised an instrument for measuring the intensity of the rays from their ability to ionize (that is, create an electrical charge in) other substances. Marie found to her surprise that unrefined uranium ore (pitchblende) was an even more potent source of the rays than was a purified uranium salt. How could that be, if the uranium in pitchblende was diluted by impurities? The answer, the Curies reasoned, must be that the ore contained *another* element—present only in a tiny proportion—that emits the rays even more intensely than uranium. "I had a passionate desire to verify this new hypothesis as rapidly as possible," Marie later wrote. They needed to find the putative element.

In seeking new elements, chemists had developed a technique called coprecipitation. Suppose a small quantity of some new element is present in a solution of another element with different chemical properties. Now you add a third element that is chemically similar to the one you're trying to isolate, then you precipitate that third element as an insoluble salt. If you're lucky, the one you're after will precipitate out too, incorporated into the same insoluble salt, due to the chemical kinship. To concentrate this trace element, you redissolve it and keep repeating the process.

When the Curies did this with pitchblende, they found that there seemed to be *two* extra sources

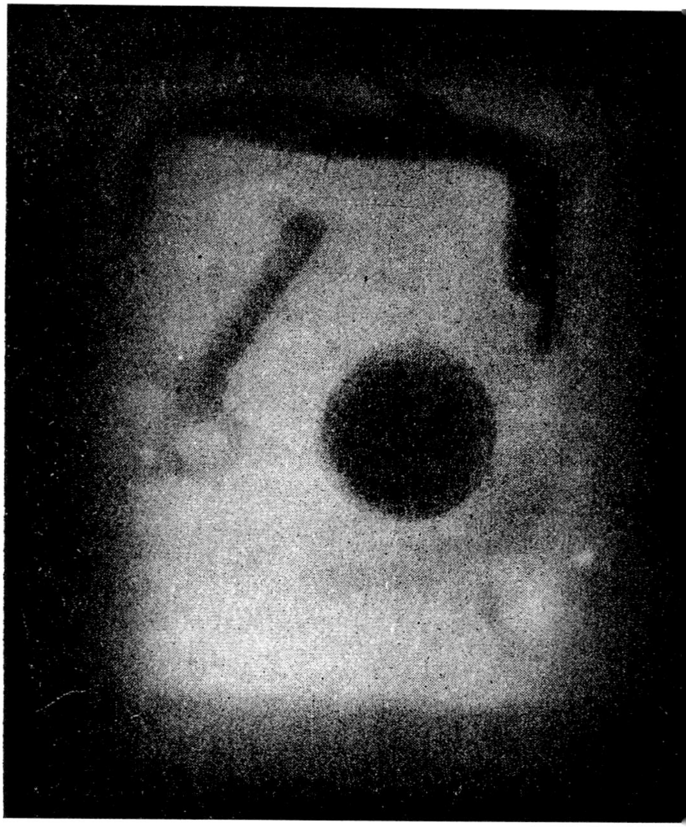

X-ray of a coin purse containing a key. From Marie Curie's *Recherches sur les Substances Radioactives*, Paris: Gauthier-Villars, 1904, Fig. 11, Wellcome Collection, London.

MARIE CURIE | 1867–1934

Marie Skłodowska-Curie, as she is now commonly known, was born in Warsaw, Poland. When she began her studies in Paris in 1891 she had very little money and faced all the prejudices against a woman trying to enter the world of science at that time; even when the 1903 Nobel prize was awarded for the discovery of radioactivity, she was almost overlooked. Yet for discovering radium and polonium, Marie also received the 1911 Nobel prize in chemistry—the first and only woman to receive two Nobels. She died in 1934 of anemia, which was believed to have been caused by her long-term exposure to radiation.

See also: Experiment 24, The nature of alpha particles and discovery of the atomic nucleus, 1908–1909 (page 108).

of uranic rays: according to their hypothesis, two new elements. One would coprecipitate with the metal barium, the other with bismuth.

The amounts were minuscule, and so to isolate enough of the two putative elements to be sure they were real, the Curies had to sift through tons of uranium ore. It was messy, laborious, and demanding labor, conducted in what was little more than a shed in the grounds of the Parisian institute. They had no fume hoods to draw off the noxious gases produced, but just had to leave the windows open when the weather was too bad to work outside. Marie did most of the work. But the effort paid off: little by little, the intensity of emission from the samples they recovered increased, as the new elements became more concentrated. In July 1898, the Curies reported that the bismuth extract contained a new element that they named *polonium*, after Marie's native country. (Nationalistic names for new elements were then all the rage.) They also proposed a name for the emission of uranic rays: radioactivity. It was, however, the barium extract that looked the most promising. Toward the end of that year, the samples were so radioactive that they glowed: the energy streaming from this other new element was visible. For that reason, in December they named it *radium*. "We were very happy in spite of the difficult conditions under which we worked," Marie wrote. "One of our joys was to go into our workroom at night, then we perceived on all sides the feebly luminous silhouettes of the bottles or capsules containing our products. It was really a lovely sight and always new to us."

To prove these samples really did contain a new element, the Curies collaborated with the French scientist Eugène Demarçay to look for a definitive fingerprint: a narrow, line-like band of absorbed light in the spectrum of the materials that didn't match any of those of the known elements. In late 1898 they finally saw such a line for radium. By the time Marie submitted her doctoral thesis in 1903, she was arguably the world expert on radioactivity. In December of that year, she became the first woman to win a Nobel

PIERRE CURIE | 1859–1906

Pierre Curie established his name in science while at the Sorbonne in Paris, where in 1880 he and his brother Paul-Jacques discovered the phenomenon of piezoelectricity. Pierre used the effect to make a quartz balance, which he and Marie utilized in their studies to measure the amount of electrical charge created by radioactivity via ionization. A physicist by training, he also made pioneering discoveries in magnetism. Pierre was tragically killed in 1906 after being struck by a horse-drawn cart on the streets of Paris.

See also: Experiment 36, The photoelectric effect, 1899–1902 (page 155).

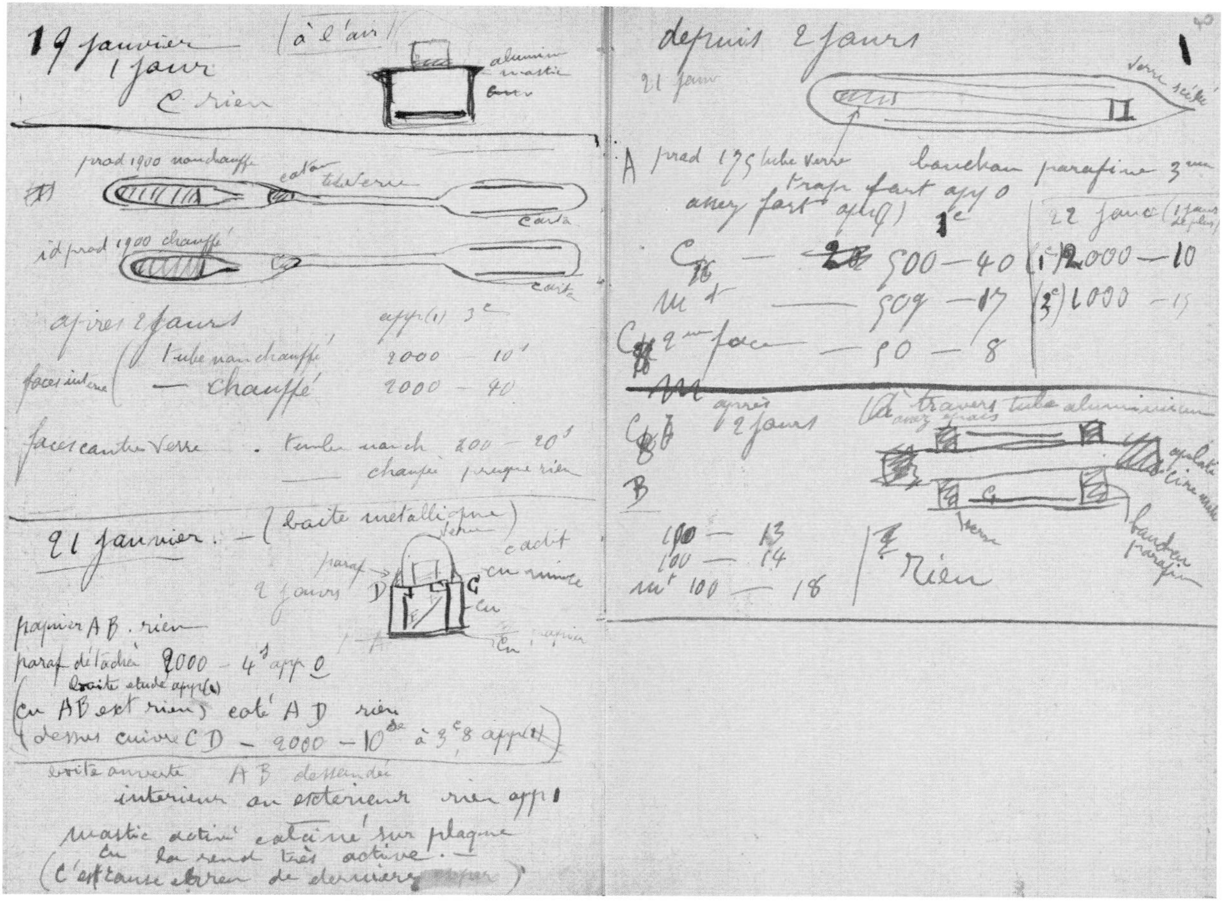

Marie Curie's holograph notebook containing notes on her experiments on radioactive substances, with sketches of apparatus, Paris, May 27, 1899 to December 4, 1902, Wellcome Collection, London. The notebook itself is contaminated with the radionuclide radium-226.

prize, being awarded the physics prize together with Pierre and Becquerel for their research on radioactivity. The Curies did not attend the Nobel ceremony, as they were too ill, afflicted with lethargy and minor ailments. Both were suffering from radiation poisoning, the unforeseen result of their research.

In 1903, the British chemist Frederick Soddy estimated how much energy must be contained inside the radium atom to produce this constant emanation that never seemed to dim. There was perhaps a million times the amount you could get from chemical reactions like those released explosively by Swedish scientist Alfred Nobel's dynamite. "The man who put his hand on the lever by which a parsimonious nature regulates so jealously the output of this store of energy," he concluded with trepidation, "could destroy the earth if he choose."

A new form of carbon (1985)

Q What kinds of molecules do carbon atoms form when they condense from a vapor?

Until 1985, we thought we knew all about carbon. This chemical element comes in two familiar forms or allotropes: hard, brilliant diamond and soft, black graphite. Both are pure carbon; the differences in appearance and properties stem from the different patterns of chemical bonds that hold the atoms together. In diamond, each carbon atom is bonded to four others in a three-dimensional network; in graphite, the atoms are joined into sheets of hexagonal rings, like chicken wire, each atom linked to three others. Weak forces hold the sheets loosely together, and their sliding gives graphite the softness that is useful for lubrication and in making pencils. But these are not the limits of carbon's versatility.

The discovery of a new form of carbon began in space. Chemist Harry Kroto at the University of Sussex, in England, was attempting to interpret the spectra of molecules in the tenuous gas that fills the space between stars. Molecules absorb electromagnetic radiation such as light and infrared at particular wavelengths, where the frequency of the molecular vibrations matches that of the light. Kroto was an expert in spectra in the microwave region, where the absorption of distant starlight is due to floppy vibrations and rotations of interstellar molecules.

Some of these molecules have a backbone made from carbon, a common element in the interstellar gas. Kroto and his colleagues wondered if some of the hitherto unexplained microwave absorption bands seen by astronomers might be caused by long, chain-like molecules of almost pure carbon, called polyynes ("polly-ines"). In 1974, Kroto and his colleagues made a polyyne in the laboratory with chains of five carbon atoms and measured its microwave spectrum. A year later, in collaboration with Canadian astronomers, they identified the same fingerprint in the spectra detected with a radio telescope in interstellar space near the center of our galaxy. By 1977 they figured they had spotted seven- and nine-carbon polyynes, too.

There seemed no obvious limit to the length of these interstellar carbon chains. But making them in the lab to compare their spectra with astronomical observations was another matter. Kroto read a paper from the 1960s by German chemists describing how they made carbon molecules with up to thirty-three atoms simply by passing a high-voltage electric current through

HARRY KROTO | 1939–2016

Born in England as Harold Walter Krotoschiner, of Silesian and German parents, Harry Kroto joined the University of Sussex in 1967 in order to work on the spectroscopy of unstable molecules. Both his interest in graphic design and his childhood passion for Meccano building sets may have helped create the strong visual sense that allowed him to discern how the C_{60} molecule is structured.

See also: Experiment 21, Building with DNA, 1991/2006 (page 96).

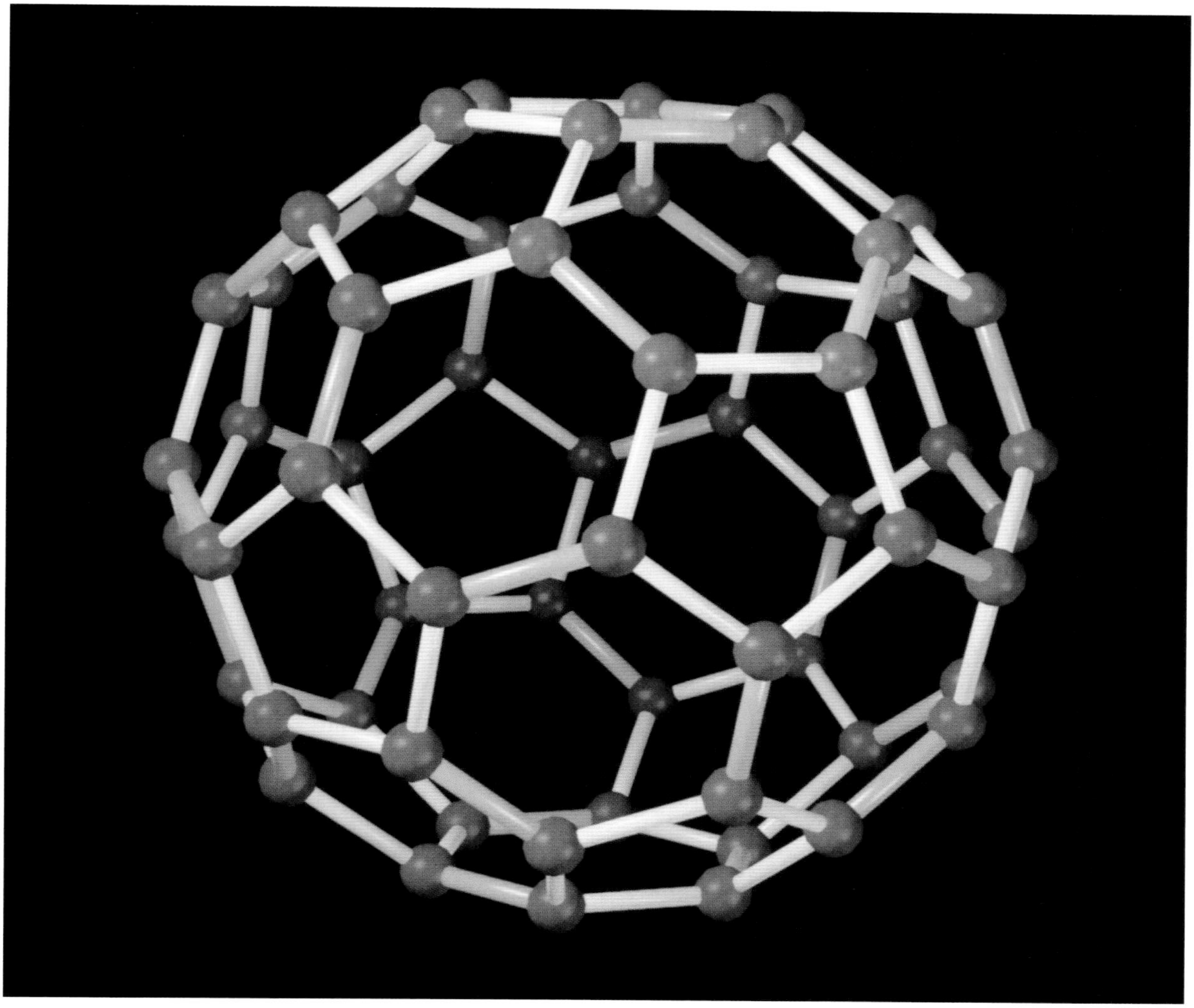

The buckminsterfullerene molecule (C_{60}). Carbon atoms are bonded into five- and six-membered rings, which fuse together to form a ball-like, closed-shell structure.

two graphite electrodes. Might that be the way to do it instead?

In 1984, Kroto was invited by US chemist Robert Curl, another microwave spectroscopist, to visit his laboratory at Rice University in Houston, Texas. There Curl was working with chemical physicist Richard Smalley on a machine that could make clusters of atoms: blobs of matter containing just a few dozen or so atoms. The researchers blasted a solid target material with an intense laser beam, vaporizing its atoms, which then gathered into clusters as they cooled while being carried along in a stream of helium. The clusters were sprayed out of a nozzle and irradiated with a second laser that knocked electrons off them and turned them into ions. In that form, their mass could be deduced using

RICHARD SMALLEY | 1943–2005

Richard Smalley is widely regarded as one of the early visionaries of the potential of nanotechnology: engineering at the scale of atoms and molecules. After discovering C_{60}, he established the Center for Nanoscale Science and Technology at Rice University, where he studied the nanometers-wide tubular structures called carbon nanotubes.

a technique called mass spectrometry, revealing how many atoms they contained.

When Kroto saw this impressive machine, he wondered if, by using a graphite target, it might produce carbon molecules like the long chains he was seeking? In fact, a team at the research labs of the petrochemicals company Exxon, in New Jersey, did precisely that experiment using a similar instrument in 1984 and reported a wide range of carbon molecules of up to one hundred or so atoms. Kroto persuaded Smalley, Curl, and their colleagues to repeat the study the following year.

Smalley was unsure they would find anything new, and anticipated spending just a week or so on the project. But his team spotted something odd: the peak in the mass spectrum corresponding to a sixty-carbon-atom molecule was often more prominent than the others. Looking back, they saw that this had been so in the Exxon work too—but by adjusting the experimental conditions, the group could make this "C_{60}" peak tower over the others like a redwood tree in an apple orchard. There was evidently something special about the C_{60} cluster.

At first, the researchers wondered if it was some layered arrangement of hexagons, like a fragment of the sheets in graphite. But Kroto had another idea. He recalled seeing one of the dome structures designed by the American architect Richard Buckminster Fuller in the 1960s, which was made of *curved* sheets of hexagons. Could the carbon sheets form a closed dome—a ball? The researchers couldn't see how to make them curl—until they heeded Kroto's recollection that Buckminster Fuller's domes might have contained pentagons too. Indeed, precisely twelve pentagons, judiciously placed so that each is surrounded by hexagons, will generate a polyhedral shell that exactly closes on itself. With twenty hexagons, the shell has a roughly spherical shape, corresponding to that of a soccer ball made from hexagonal and pentagonal patches. At each corner of the polygons, where three edges meet, there sits a carbon atom—sixty of them in the entire shell.

Kroto, Rice, and colleagues lacked definitive proof that this was the structure of their C_{60} clusters. But the idea seemed so beautiful that it just had to be right: the special stability of this closed shell would explain why it so dominated the mass spectrum. In 1985, they presented their results and interpretation in a paper in the journal *Nature*, where they proposed a name for the C_{60} molecule: buckminsterfullerene.

It soon became clear that a whole family of closed carbon shells was possible, each with twelve pentagons but different numbers of hexagons; these became known as the fullerenes. None, however, rivaled the elegant symmetry

ROBERT CURL | 1933–2022

Robert Curl, an expert in the technique of microwave spectroscopy, was often regarded as the quietest and most modest member of the Nobel triumvirate who discovered C_{60}. When the President of Rice University asked what he could do for Curl after the Nobel announcement, Curl answered that it would be nice to have a bicycle rack installed closer to his office and laboratory. He always asserted that the two graduate students James Heath and Sean O'Brien, who made the C_{60} experiments work, deserved an equal share of the recognition.

This rock from the Sudbury meteorite impact crater in Ontario, Canada (formed 1.85 billion years ago), contains C_{60} molecules thought to have been produced in the plume of impact debris from carbon-containing material in the meteorite.

of C_{60}. It was not until five years later that other researchers found a way to produce fullerenes in large quantities by vaporizing graphite electrodes, enabling them to be purified and crystallized so that their structure could be confirmed by X-ray crystallography. Soon after, that same method was used to make other large carbon molecules from curled-up, graphite-like sheets, in particular tubes with hemispherical half-fullerene end caps, just a few nanometers (millionths of a millimeter) across. The age of "carbon nanotechnology" had arrived. Smalley, Curl, and Kroto were awarded the 1996 Nobel prize in chemistry for their work.

But what about those carbon molecules found in space? It is now clear that fullerenes form there too; they have even been discovered in carbon-rich meteorites.

Building with DNA (1991/2006)

Q Can we use DNA as a programmable construction material?

The discovery of the structure of the DNA molecule by James Watson and Francis Crick in 1953 showed how it could encode information in chemical form: as a sequence of the four component "letters" (the nucleotide bases) strung together like beads on a thread. While this discovery launched the field of modern molecular genetics and gene-sequencing, it was almost three decades later that researchers realized DNA's information-bearing nature also made it a potential construction material that could be programmed to assemble into specific shapes.

Since the 1970s, chemists had been pondering how to design molecules that would assemble themselves spontaneously into larger structures. This happens during the formation of crystals: the atoms or molecules stack into well-ordered arrays. But molecular scientists had little control over those arrangements. Molecular self-assembly also happens in living cells, where proteins and other molecules gather together into complex arrangements with biological functions. Those arrangements do not generally have crystalline regularity, but neither are they arbitrary: they are determined by the shapes and properties of the molecules themselves, which fit together like jigsaw pieces. If chemists could master this art of molecular self-assembly, they figured they could make molecular-scale structures by design, perhaps creating new materials or tiny machines. The field became known as supramolecular chemistry: chemistry beyond the molecule.

In the 1970s, American biochemist Nadrian ("Ned") Seeman learned the art of X-ray crystallography, which can reveal the shapes of the molecule. At the Massachusetts Institute of Technology (MIT) he studied the chemistry of DNA under the guidance of Alexander Rich, who had collaborated with Francis Crick. While subsequently working at the University of New York at Albany, Seeman was approached by a colleague who asked him if he knew how to put DNA strands together in an arrangement in

NADRIAN SEEMAN | 1945–2021

Born in Chicago in 1945, Nadrian Seeman was a self-professed "Sputnik kid" who was amazed to discover as a medical undergraduate at the University of Chicago that if he became a professor, "I could spend most of my day having fun by doing research and also get paid to do it." In the event, academia proved less fun at first than he had imagined: he drifted, dissatisfied, from one position to another, until hit by the inspiration for building with DNA. Seeman was renowned for plain speaking: he was, in Paul Rothemund's words, "a singular character … at once gruff and caring, vulgar and articulate, stubborn and visionary." Largely ignored at first, Seeman's pioneering work eventually brought him many honors and awards, including the Kavli Prize in Nanoscience in 2010.

See also: Experiment 18, The three-dimensional shape of sugar molecules, 1891 (page 84); Experiment 49, The proof that DNA is the genetic material, 1951–1952 (page 197).

which three strands entwine to make a branching junction, like a Y-shape. These so-called Holliday junctions can sometimes occur in nature.

As Seeman sat pondering the problem in a bar, he recalled a picture by Dutch geometric artist M. C. Escher in which stylized fish were linked via their fins and tails into a three-dimensional grid. He realized that the intersections of the grids were like Holliday junctions. If he could make DNA Holliday junctions and somehow give them sticky ends, they might link into a crystalline array like the one Escher drew. "Thinking of the fish as nucleic acids, I imagined their contacts being programmed by sticky-ended cohesion," Seeman said. "This was my major epiphany."

To give the DNA double-helices sticky ends, all he had to do was to make one strand a bit longer than the other, so that a short stretch of it was left dangling and unpaired. The strands stick together by the pairing up of nucleotide bases when their sequences match: each base has a complementary partner that it prefers to stick to. So by choosing the right sequences, the DNA junctions could be programmed to assemble to order. Chemists had by that time devised methods for synthesizing DNA strands with specified sequences. One might then just mix the strands in a test tube and let them assemble themselves in the right way. In 1982 Seeman published a paper outlining the notion of making lattices from DNA junctions,

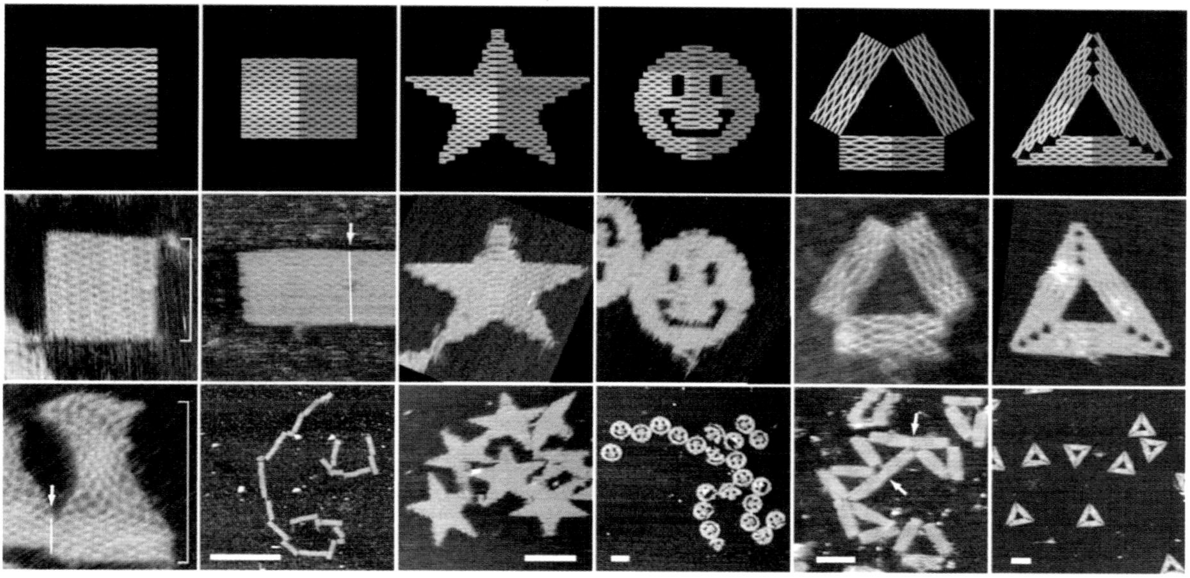

DNA nanotechnology: the origami structures made by Paul Rothemund in 2006 from DNA molecules designed to fold into specific shapes. The top frames show the theoretical target structures, while the lower ones show the experimental shapes at different scales of magnification. Each shape is typically about 100 nanometers across. From Paul W. K. Rothemund's "Folding DNA to create nanoscale shapes and patterns," *Nature*, March 16, 2006, Vol. 440,16, Fig. 1.

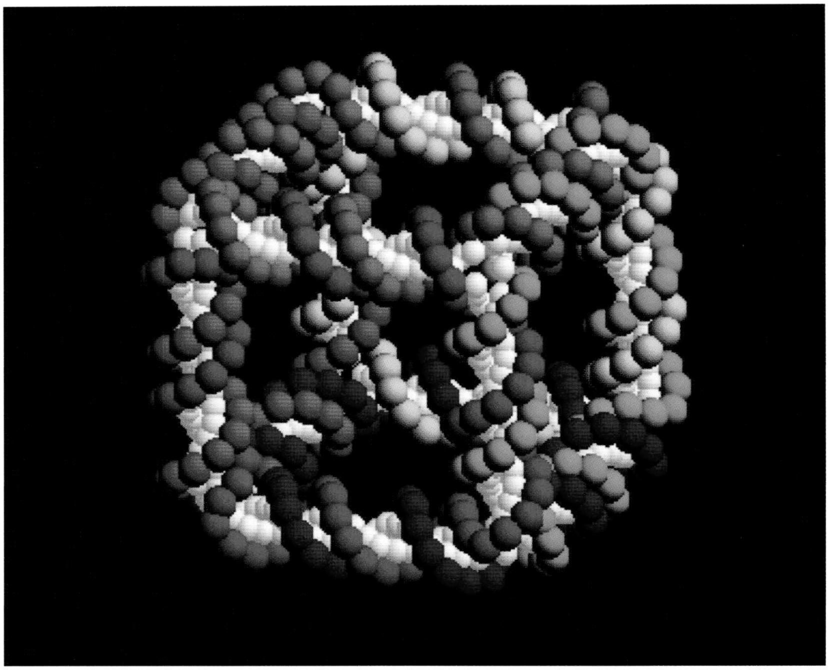

To observe the structure of the molecule in Ned Seeman's DNA cube, follow the red strand around its cycle: It is linked twice to the green strand, twice to the cyan strand, twice to the magenta strand, and twice to the dark blue strand. The same pattern of linkages is found for each of these strands, and it weaves them into a three-dimensional shape that is roughly cubical. The Holliday junctions between three strands of DNA can be seen at each corner of the cube. Nadrian C. Seeman, New York University, 1991.

but he said it got "zero reaction." It was too far ahead of its time: no one else had realized that DNA might have uses beyond biology.

Seeman reasoned that the way to make people take notice was to prove the idea experimentally. It wasn't until 1991, by which time he had moved to the chemistry department of New York University, that he was able to do this. He and his graduate student Junghuei Chen designed and made DNA strands that they thought would assemble into Holliday junctions, which would then stick together as the corners of a cube. The whole structure would be no bigger than a typical protein molecule.

Making sure the assembly worked as planned was no mean feat: the researchers had to assemble the strands into the cube faces bit by bit, purifying the products at each step before going onto the next. But it worked. When the two announced their success in *Nature*, the work was regarded as something of a baffling curiosity. Seeman subsequently produced more ambitious polyhedral DNA shapes, as well as mesh-like, two-dimensional DNA sheets. He and others realized that the assembly process could be made algorithmic: the programmed assembly of DNA "tiles" could be used to carry out a computation, in which the tiles encode the problem and the way they assemble reveals the answer.

In 2006 the algorithmic approach to DNA self-assembly was extended by Paul Rothemund at the California Institute of Technology (Caltech), who had trained in both biology and computer science. Rothemund showed that he could decide on some target shape—it could be a molecular-scale smiley face or a map of the world—and then use a computer program to work out what the sequence of a DNA strand had to be, so it folded up into that shape by the pairing of different regions of the strand. It was a kind of DNA origami. Others have designed DNA strands that will pair and unzip repeatedly to produce machine-like motion. Seeman's insight

has now led to an entire discipline of DNA nanotechnology: a technology conducted at the molecular scale of nanometers. "We don't have to have all the ideas anymore, and we don't have to make all the mistakes," said Seeman in 2010. "The field has been launched."

M. C. Escher's wood engraving and woodcut, *Depth*, October 1955, The M. C. Escher Company, Baarn, the Netherlands. This image served as the inspiration for Ned Seeman's idea of making geometric shapes from branching junctions of double-stranded DNA.

Looking into matter

To explain the world, science makes a bold gambit: it invokes entities that are not directly visible or perceptible. The Greek philosopher Anaximander in the sixth century BC was one of the first to do so, positing a kind of primal substance called *aperion*. About a hundred years later, Leucippus and Democritus went further by arguing that all material things are divided into particles—atoms—too tiny to see. Just how fine-grained, then, is the stuff of reality? In the seventeenth century, some hoped that the microscope might reveal this grain, disclosing—as the physician Henry Power put it—"the solary atoms of light [and] the springy particles of air." He did not reckon on just how small these putative atoms were, and it was not until the twentieth century that experimental methods acquired the extraordinary acuity needed to reveal them—and eventually, even to manipulate them, one by one.

First studies with the microscope (1625-1665)

Q What does the world look like very close up?

Magnifying lenses have a long history. They are described by the Arab scholar Abū ʿAlī al-Ḥasan ibn al-Haytham (see page 137) in the eleventh century, and the Franciscan friar Roger Bacon—commonly (if rather meaninglessly) described as the first experimental scientist—wondered in 1266 if the lens's ability to make objects look bigger might help elderly people with weak eyes. That idea seems to have been put into practice soon after; in 1306 a Florentine cleric commented: "It is not yet twenty years since there was found the art of making eyeglasses." At this time the best lenses were made of quartz, but during the fourteenth century the skills of glassmakers and lens-grinders advanced to the point where cheaper glass lenses could do a good job instead.

With the assistance of magnifying lenses, Bacon wrote, we might "number the smallest particles of dust and sand." Yet it was not until another three hundred and more years had passed that the methodical study of the microscopic world took off. Why the delay, given that the materials and techniques used by the first microscopists in the early seventeenth century were mostly available during the Middle Ages? Perhaps part of the answer is that previously no one believed there was anything much to be found there. Why would God make anything too small for the human eye to see?

Many historians credit the first microscope to the Dutch maker of eyeglasses, Hans Janssen of Middelburg, and his son Zacharias, in around 1590. According to a description given in 1650 by the diplomat William Boreel, who knew the Janssen family, their microscope was a cumbersome instrument: a tube about 2½ feet long, held vertically in a brass tripod, with a lens at each end: a version of the familiar double-lens instrument used in schools today, and which apparently provided magnification by a factor of 3–9. An instrument made by the Janssens and dated 1595 is, however, more portable: handheld, with tubes that slide inside one another.

Other sources say that the invention was made around the same time by Hans Lippershey, also of Middelburg, who is similarly credited with inventing the telescope—an idea that Galileo quickly copied the moment he heard about it. At any rate, Galileo seems to have got hold of a microscope too by the 1620s. In 1624, he wrote to his rich young patron Federico Cesi that "I have observed many tiny animals with great admiration, among which the flea is quite horrible, the mosquito and the moth very beautiful." In 1625, Galileo's colleague Francesco Stelluti in the exclusive little experimental club in Italy called

Telescope measuring 36½ inches in length made by Galileo from wood and leather in late 1609 to early 1610, Museo Galileo, Florence, Italy.

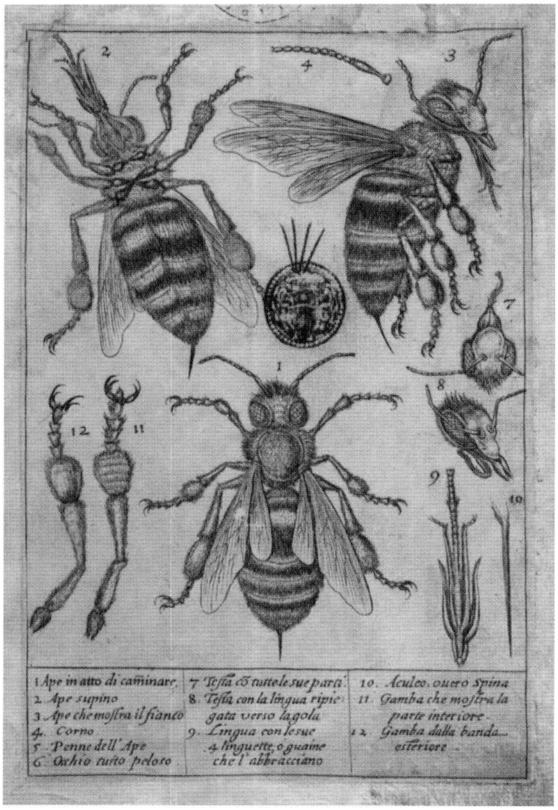

Magnified illustration of bees. From Francesco Stelluti's *Apiarium*, Rome: 1625, Museo Galileo, Florence, Italy.

the Accademia dei Lincei (Academy of Lynxes) published a text on bees accompanied by a marvelous engraving showing the details of these insects he had observed with the *microscopio*.

Yet natural philosophers did little more than dabble with the device for a few decades. The English physician Henry Power's book *Experimental Philosophy* popularized the instrument in 1664, prompting the curious and wealthy (including the diarist Samuel Pepys) to go and buy one, at considerable expense, from instrument-makers happy to profit from the new craze.

Yet it was the book published the following year—*Micrographia* by Robert Hooke—that changed everything. In one sense, Hooke's observations barely qualify as an experiment at all: he simply looked at lots of different objects using the devices created for him by the London instrument-maker Christopher Cock, which used a water-filled globe to focus the light of a lamp onto the specimen. However, Hooke presented his studies as an exciting journey into a microcosm of unimaginable marvels: "new Worlds and Terra Incognitas to our view," as he put it. Curious experimenters, Hooke explained, were not merely using the microscope's lenses to see things more accurately; they were also able to see entirely novel aspects of how nature works. They were engaging in scientific discovery.

Micrographia set out this case with stunning persuasiveness. In large-format pages, some folding out to even greater size, Hooke presented breathtaking drawings of what he had seen: images that he drew himself. Here were fantastical landscapes made from mold, the tip of a needle revealed in all its scratchy imperfection, the scaly armor of a gigantic flea, and the multifaceted eyes of a fly, in each hemispheric facet of which Hooke

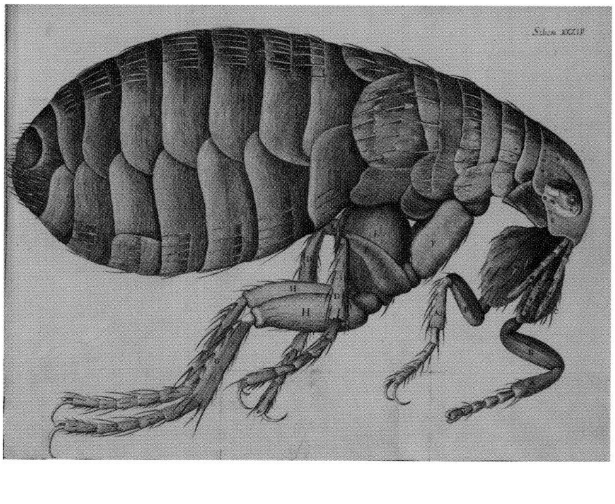

Engraving of a flea. From Robert Hooke's *Micrographia*, London: Printed by J. Martyn and J. Allestry, 1665, Plate XXXIV, Wellcome Collection, London.

ROBERT HOOKE | 1635–1703

Born on the Isle of Wight, Robert Hooke was a key figure in the early Royal Society, which was formed in London in 1650 by a group of men enthusiastic about the new "experimental philosophy." Hooke was charged with providing "demonstrations" at the Society's gatherings, thanks to his skills in devising and conducting experiments. His interests were wide-ranging, including astronomy, optics, mechanics (students today still learn Hooke's law of springs), and architecture—with Christopher Wren he devised a plan to rebuild London after the Great Fire of 1666. A sometimes irascible man, Hooke earned the enmity of Isaac Newton, who is rumored (perhaps apocryphally) to have ensured that no portraits of Hooke survived at the Royal Society after his death. At any rate, none is known.

See also: Experiment 31, The camera obscura, early 11th century (page 137); Experiment 41, The microscopic observations of microbes, 1670s (page 170).

Robert Hooke's compound microscope and its illuminating system, 1665–1675, Science Museum, London.

claimed (a trifle improbably) to be able to make out the reflection of a tree through his window. In this intricacy of the natural world, Hooke discerned theological as well as scientific implications. "There may be as much curiosity of contrivance in every one of these Pearls, as in the eye of a Whale or Elephant, and the almighty's *Fiat* could as easily cause the existence of the one as the other." As Power had put it in verse, "God is greatest in the least of things/And in the smallest print we gather hence/The World may Best reade his omnipotence."

There was, it seemed, nothing so small and insignificant that it was not worth studying with the microscope, and in its invisibly small details experimenters might hope to discern the mechanisms of the natural world.

FRANCESCO STELLUTI 1577–1652

Little known today, Francesco Stelluti exemplifies the notion of the Renaissance man. A lawyer by training, he made contributions to literary studies, cartography, mathematics, and science. With his aristocratic patron Federico Cesi and two other like-minded colleagues, he founded the *Accademia dei Lincei* in 1603, dedicated to the close study of the natural world—the name refers to the lynx's legendarily keen eyesight. In 1611, the little "academy" recruited a new member: Galileo.

Understanding Brownian motion (1908)

Q Can we prove that atoms are real, even if they are too small to see?

In 1828, the eminent English botanist Robert Brown reported that tiny particles suspended in water, observed through a microscope, seemed to execute a frenetic and unceasing dance. He called them (somewhat misleadingly, as others soon pointed out) "active molecules."

In some ways, this was nothing new. As Brown admitted, microscopic observations going back to Antonie van Leeuwenhoek in the seventeenth century had revealed invisibly tiny entities that swim around in water. But Brown's findings showed that this movement could have nothing to do with life—for he saw it not just in pollen grains but also in specks of non-living matter such as pieces of ground-up window glass and even, bizarrely, the dusty debris from a piece of the ancient Egyptian Sphinx.

What physical forces could account for this motion, which later became known as "Brownian"? Might it be due to electrical repulsion between particles, or some effect of surface tension? One prominent idea was that Brownian motion arose from the agitation caused by heat, much as the theories of James Clerk Maxwell and Ludwig Boltzmann in the latter half of the century seemed to imply. Their "kinetic theory" of gases suggested that all molecules in gases (and liquids) were constantly jiggling because of their heat energy—that heat was, indeed, the result of mere molecular motion.

However, even the very existence of atoms and molecules was not universally accepted. As there was no direct evidence for them, some scientists insisted that the "atomic hypothesis" had to be regarded as just that: a convenient idea, but not necessarily true. Among those who believed in the reality of atoms and molecules was Albert Einstein, and in 1905 he published a paper arguing that this could explain Brownian motion. Einstein assumed that the jiggling of suspended particles was caused by the impacts of innumerable molecules in the surrounding liquid. Tiny, chance imbalances in the number of impacts on different sides of a particle could be enough to push it randomly in one direction or another, so that it executes a meandering, erratic path through the liquid.

Einstein calculated what this would imply for the nature of Brownian motion. He showed that the average distance such a particle moved from its initial position over some fixed time interval depended on the square root of that timespan. Here was a prediction that experimentalists could test—something they had lacked for almost eighty years since Brown's observations. As historian of science Stephen Brush has said, "three-quarters of a century of experimentation produced almost no useful results, simply because no theorist [before Einstein] had told experimentalists what quantity should be measured!"

On seeing Einstein's work, the Polish mathematician Marian von Smoluchowski published a paper the following year saying that Einstein's results "agree completely with those that I obtained several years ago by following a completely different line of thought"—one, he could not resist adding, that was "more direct, simpler, and more convincing." It wasn't just sour grapes—Smoluchowski had a point in suggesting that Einstein's reasoning was a bit obscure. But it was Einstein's work that prompted the French physical chemist Jean-Baptiste Perrin to make

Clarkia pulchella, or ragged robin, the pollen of which was the basis for the discovery of Brownian motion. From Frederick Pursh's *Flora Americae Septentrionalis* ... London: Printed for White, Cochrane, and Co., 1814, Vol. 1, Plate II, Missouri Botanical Garden.

measurements of Brownian motion that might prove the existence of atoms and molecules (a minority view in France at that time). Einstein's paper gave Perrin what he was looking for: as he put it, "a crucial experiment that, by approaching the molecular scale, might give a solid experimental basis to attack or defend the kinetic theories"—and their assumption of the reality of molecules.

Perrin was a specialist in colloid chemistry—loosely speaking, this is the science of small, suspended particles—at the Sorbonne in Paris. With the assistance of his student M. Dabrowski, in 1908 he devised a way to track the motion of individual colloidal particles such as tiny droplets of the sticky resin called mastic that could be extracted from the bark of certain trees. This involved sandwiching a very thin glass sheet

JEAN-BAPTISTE PERRIN
1870–1942

Like many physicists in the late nineteenth century, Jean-Baptiste Perrin was initially drawn to the study of cathode rays and X-rays. After his studies of Brownian motion in the 1900s, he studied nuclear chemistry, correctly proposing in 1919 that stars are fueled by the fusion of hydrogen atoms. In the 1930s he petitioned the French government to set up a framework for funding and supporting scientific research, which became the *Centre National de la Recherche Scientifique* (CNRS), France's key research agency to this day.

See also: Experiment 11, The origin of heat, 1847 (page 52); Experiment 26, Invention of the scanning tunneling microscope and moving single atoms, 1981–1982/1989 (page 116).

(about 0.1 millimeters thick), in which a hole had been bored, between two glass slides to create a small cylindrical cavity in which the droplets could be individually observed in a microscope. Perrin engraved a square grid on the glass side to track the particle movements, and he used a device called a camera lucida to project the image from the microscope onto a sheet of white paper where he or Dabrowski could draw the droplet positions at regular time intervals. Other researchers were attempting similar investigations using photographic techniques, which some asserted were more objective and therefore more reliable. But photographic emulsion wasn't always sensitive enough to register a light liquid droplet against a light background—the eye, Perrin said, was more dependable. He used a centrifuge to separate particles of different sizes so that he could experiment on ones all the same size: typically about 0.1 thousandths of a millimeter.

It required immense patience and diligence to follow and plot the trajectories of single particles over many minutes. Perrin was not an obvious candidate for such work—the French scientist Louis de Broglie later pronounced him "quite distracted, of a rather impulsive character, such that one might have thought him little suited to carry out a task requiring so much attention and perseverance."

Be that as it may, Perrin did persist, and in 1909 he published the fruits of his studies in a work titled "Brownian movement and molecular reality." Here he presented the jagged, wandering paths he had observed for his particles, and he confirmed that their motion fitted Einstein's square-root law for the relationship between displacement and time. "I think," he wrote, "that it will henceforth be difficult to defend by rational arguments a hostile attitude to molecular hypotheses." Einstein was delighted, admitting to Perrin that "I would have thought such a precise study of Brownian motion impossible to realize."

Most other scientists were persuaded. "After this," wrote the eminent Swedish chemist Svante Arrhenius in 1911, "it does not seem possible to doubt that the molecular theory entertained by the philosophers of antiquity, Leucippus and Democritus, has attained the truth, at least in essentials." Perrin's findings even convinced the German physical chemist Wilhelm Ostwald, a long-standing sceptic of atomism. In 1926, Perrin was awarded the Nobel prize in physics for showing that molecules are real after all.

Microscopic observations of particles suspended in water: left, gamboge yellow pigment; right, mastic, a resin produced by the evergreen tree *Pistacia lentiscus*). From Jean Perrin's *Brownian Movement and Molecular Reality*, London: Taylor and Francis, 1910, Snell Library, Northeastern University, Boston, Massachusetts.

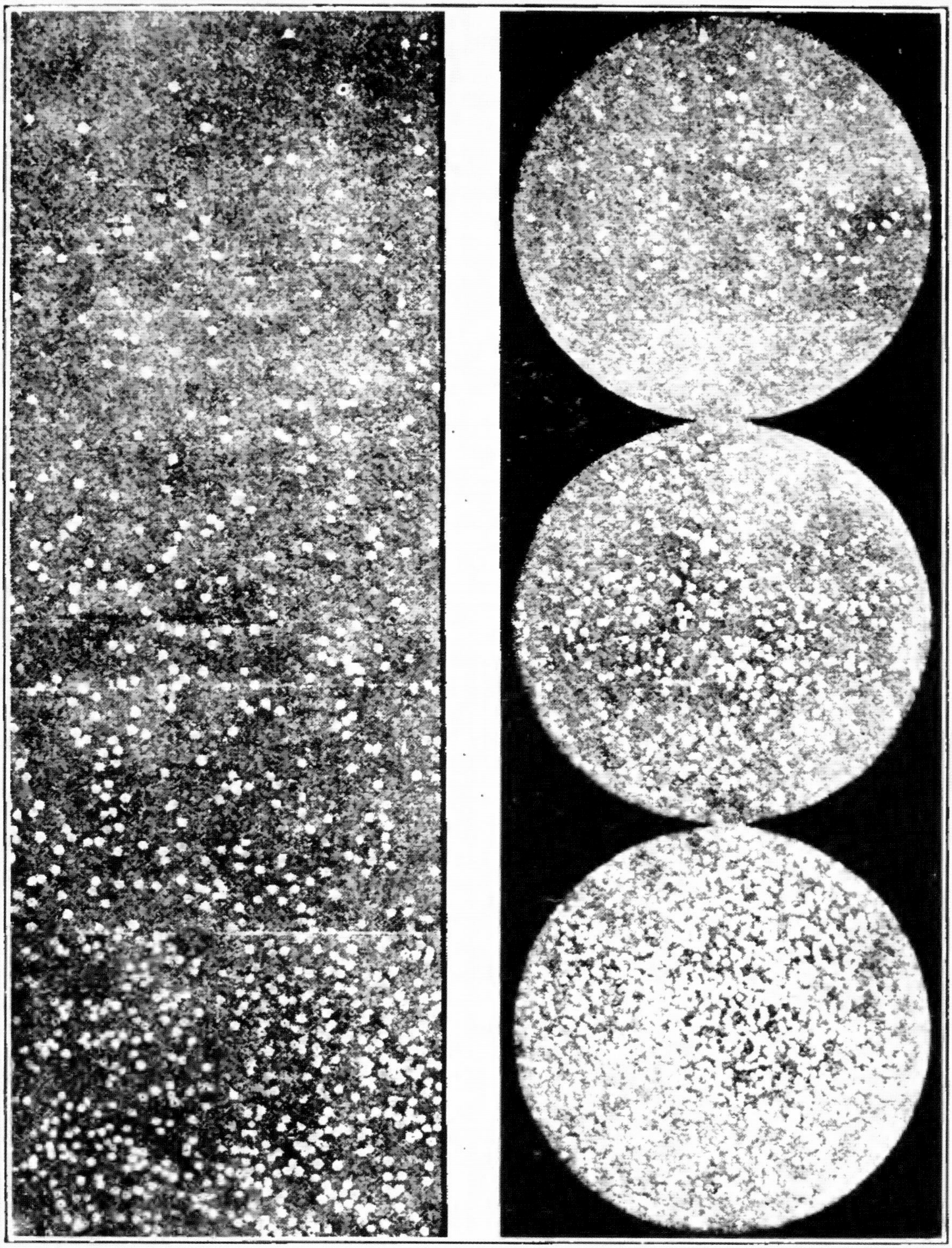

The nature of alpha particles and discovery of the atomic nucleus (1908-1909)

Q What are the alpha particles emitted in radioactive decay? What is the internal structure of atoms?

No individual better illustrates how science advances through ingenious, well-conceived experiments than the New Zealand-born physicist Ernest Rutherford. He was a force of nature: with a lack of airs that his colleagues attributed to his antipodean upbringing, he was given to marching around his lab singing (badly) "Onward Christian Soldiers"—truly a "tribal chief," according to one of his students. "He didn't look in the least like an intellectual," wrote the scientist and novelist C. P. Snow. "But no one could have enjoyed himself more, either in creative work or the honors it brought him. He worked hard, but with immense gusto."

It is hard to believe with what primitive means Rutherford probed the innermost secrets of the atom in the early twentieth century. Famously, he had no time for elaborate, expensive equipment, preferring the philosophy attributed to the

Ernest Rutherford (right) and Hans Geiger with their apparatus for counting alpha particles, 1912, Science Museum, London.

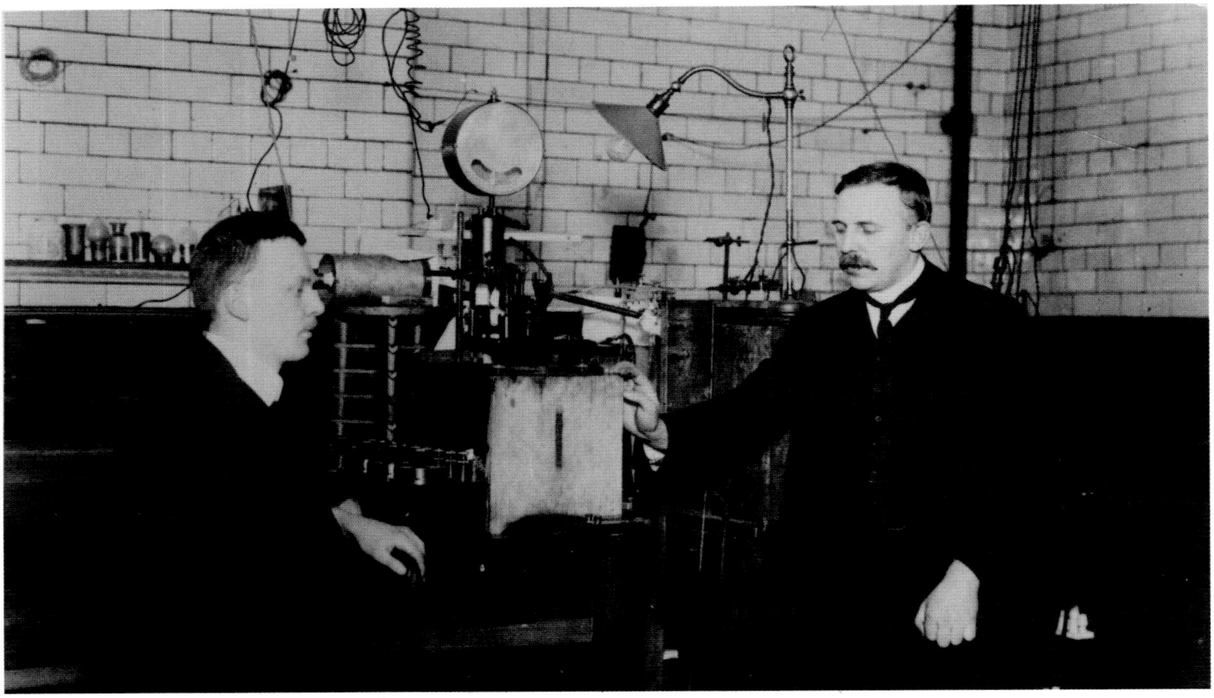

Cavendish Laboratory at Cambridge, where he worked in the 1920s and '30s, of improvising with "sealing wax and string." With such simple means, Rutherford clarified what the Curies' radioactivity is, deduced the structure of the atom, and, in 1919, reported the first instance of induced radioactive decay: in the popular phrase of the moment, "splitting the atom." His work earned him a Nobel prize in 1908, and he is commemorated today in the name of the artificial superheavy chemical element rutherfordium.

In the late 1890s, Rutherford deduced that the "rays" emanating from radioactive elements such as uranium are of at least two types. Aluminum foil would block some of the emission from uranium, but would have no influence on the remainder. He called the more easily absorbed rays "alpha radiation" and the more penetrating "beta radiation." A third, even more penetrating form was later discovered: gamma radiation.

What were these rays? Rutherford set out to understand alpha rays and, working at McGill University in 1900–1903 with chemist Frederick Soddy, he showed that they would be deflected in their course by electric fields, suggesting that they were actually electrically charged particles with a mass comparable to that of a hydrogen atom. In other words, when an atom decayed by emitting alpha particles it lost a bit of its mass, turning it into another kind of atom entirely. Thus, in 1902, Rutherford proposed that radioactive decay caused the transmutation of one element to another.

Rutherford and Soddy couldn't measure the charge and mass of alpha particles, but only their ratio. This pointed to two possibilities. Either the particles had a charge equal and opposite to that on a single electron (which is negatively charged), and twice the mass of a hydrogen atom, or two electron charges and four times the mass of hydrogen, giving them the same mass as a helium atom. Rutherford suspected the latter, and when he left McGill in 1907 to work at Manchester University in England, he tested that hypothesis.

Rutherford's student at Manchester, Hans Geiger, devised an instrument for detecting alpha particles one by one. Using this device, in 1908 Rutherford and Geiger were able to measure how much charge a given number of alpha particles carried, and thus deduce their absolute charge—which indeed seemed to be twice the electron charge. "An alpha-particle," they pronounced, "is a helium atom" (more properly, a positively charged helium atom, or ion). Still, Rutherford set out to gather definitive proof in an experiment of unsurpassed elegance. If alpha particles were, indeed, essentially helium ions, then their accumulation would produce a small amount of helium gas—which could be identified as such from the distinctive wavelength of light it emitted when stimulated by passing an electrical current through it.

The experiment was classic Rutherford: simplicity of design coupled with irrefutability of the result. He commissioned a local glassblower to make a small glass tube with walls so thin that alpha particles from a radioactive source (radium) inside the tube could pass right through them. This tube was enclosed in a larger, stouter one, so that the alpha particles would accumulate in the outer tube as—if the hypothesis was correct—helium gas.

Schematic illustration of Ernest Rutherford's famous gold-foil experiment, in which he deduced that atoms have very dense nuclei, much smaller than the atoms themselves, surrounded by electrons.

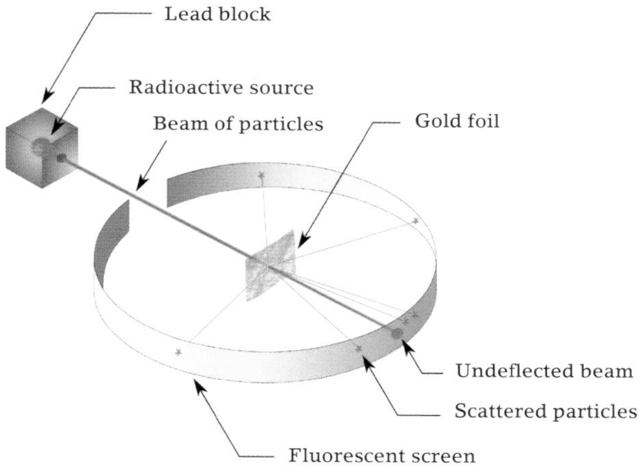

ERNEST RUTHERFORD
1871–1937

No one contributed more to the understanding of the structure of the atom and the inception of the nuclear age than Ernest Rutherford. He came to England in 1895 to work at the Cavendish Laboratory in Cambridge under J. J. Thomson. He later studied at McGill University in Canada and at Manchester University, England, before becoming director of the Cavendish in 1919. There he oversaw the discovery of the neutron and the invention of the particle accelerator. After a career that included a Nobel prize, presidency of the Royal Society, and a knighthood, he was buried in Westminster Abbey alongside Isaac Newton and Charles Darwin.

See also: Experiment 25, Measuring the charge on an electron, 1909–1913 (page 112); Experiment 27, The cloud chamber, 1894–1911 (page 121).

An electrical discharge at the top of the tube would then excite the helium to emit its telltale glow. Working with his student Thomas Royds, in 1908 Rutherford found his proof. In his Nobel lecture that year, he speculated that "other elements may be built up in part from helium."

Built how? How were the elementary particles arranged inside atoms? As well as investigating what alpha particles are, Rutherford realized that they could be used as a probe for studying atoms. They are ejected with considerable energy, and he figured that by firing them like bullets at other atoms he might be able to deduce from the way they are scattered what the inside of atoms is like. In 1909, with Geiger and fellow student Ernest Marsden at Manchester, Rutherford shot alpha particles at thin foils of metals like aluminum, silver, and gold, and measured how they were scattered. They found that most of the particles would pass straight through, suggesting that atoms weren't simply dense spheres packed together but were more tenuous, perhaps mostly empty space. Occasionally, the particle projectiles would be slightly deflected in their course by bouncing off the atoms in the foil.

Rutherford's intuition led him to suggest to his students that they look for particles that were reflected straight back from the foils. This seemed unlikely; the energetic particles were expected just to punch their way through the ultra-thin metal. Yet Geiger and Marsden found to their astonishment that a few alpha particles did indeed bounce right back. It was, Rutherford said, "as if you had fired a 15-inch shell at a piece of tissue paper and it came back and hit you … [It was] quite the most incredible event that has ever happened to me in my life."

Rutherford figured that the only way this could happen is if nearly all the mass of an atom is concentrated in an incredibly dense, positively charged nucleus at its center, surrounded by empty space in which the negatively charged electrons orbit. This "nuclear model" of the atom was somewhat similar to one proposed in 1903 by Japanese physicist Hantaro Nagaoka. The corollary was that, when a radioactive atom decays by emitting an alpha particle, it loses a fragment of its nucleus.

The idea—like most good ideas in science—raised as many questions as it answered. What, for example, would keep a negatively charged electron orbiting a positive nucleus, rather than spiraling inward because of the electrical attraction between the two? An answer was proposed in 1912 by the Danish physicist Niels Bohr, based on Einstein's proposal in 1905 that energies at the atomic scale are "quantized": they can only take certain values, like rungs on a ladder. Rutherford's atom marked not just the beginning of the nuclear age, but a key stage in the quantum revolution in physics.

Ernest Rutherford's notes on the "solar system" model of the atom, ca. 1908, Cavendish Laboratory, Cambridge.

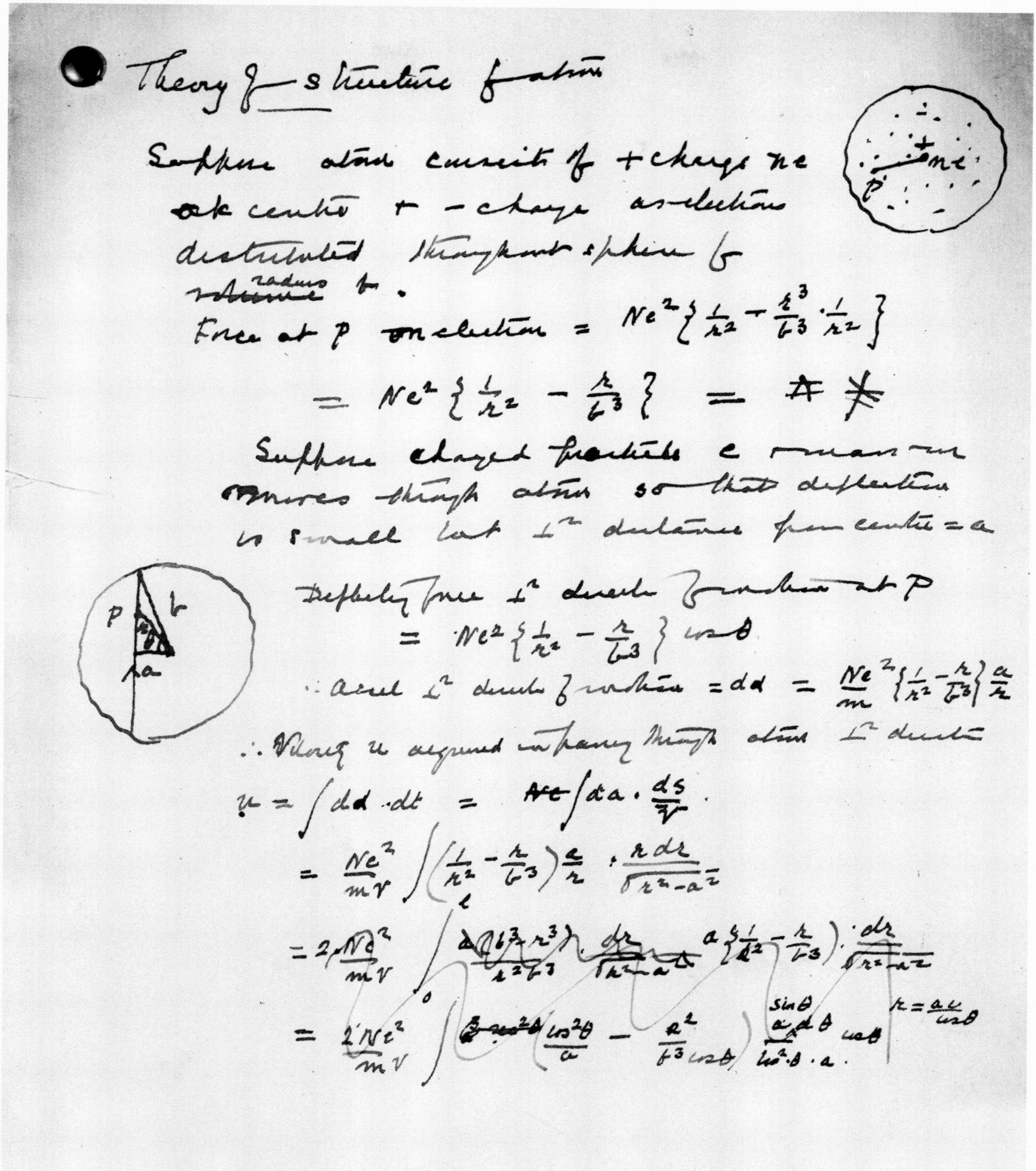
Theory of structure of atom

Suppose atom consist of + charge Ne at centre + − charge as electrons distributed throughout sphere of radius b.

Force at P on electron $= Ne^2\left\{\frac{1}{r^2} - \frac{r^3}{b^3}\cdot\frac{1}{r^2}\right\}$

$= Ne^2\left\{\frac{1}{r^2} - \frac{r}{b^3}\right\} = A$

Suppose charged particle e mass m moves through atom so that deflection is small but ⊥ distance from centre = a

Deflecting force ⊥ direction of motion at P

$= Ne^2\left\{\frac{1}{r^2} - \frac{r}{b^3}\right\}\cos\theta$

∴ accel ⊥ direction of motion $= d\alpha = \frac{Ne^2}{m}\left\{\frac{1}{r^2} - \frac{r}{b^3}\right\}\frac{a}{r}$

∴ Velocity u acquired in passing through atom ⊥ direction

$u = \int d\alpha \cdot dt = \int d\alpha \cdot \frac{ds}{V}$

$= \frac{Ne^2}{mV}\int \left(\frac{1}{r^2} - \frac{r}{b^3}\right)\frac{a}{r}\cdot\frac{r\,dr}{\sqrt{r^2-a^2}}$

$= 2\frac{Ne^2}{mV}\int_0 a\left\{\frac{1}{r^2} - \frac{r}{b^3}\right\}\cdot\frac{dr}{\sqrt{r^2-a^2}}$

$= \frac{2Ne^2}{mV}\int\left(\frac{\cos^2\theta}{a} - \frac{a^2}{b^3\cos\theta}\right)\frac{\sin\theta\, a\, d\theta\, \cos\theta}{\cos^2\theta\cdot a}$ $\quad r = \frac{a}{\cos\theta}$

Measuring the charge on an electron (1909-1913)

Q What is the smallest unit of electrical charge?

In 1897, Joseph John ("J. J.") Thomson, working in the Cavendish Laboratory at the University of Cambridge, found the first subatomic particle: the electron, the lightweight entity that was later revealed to surround the dense atomic nucleus. It was already known that the mysterious "cathode rays" emanating from metal electrodes could be bent by magnetic fields, suggesting that they were composed of particles with negative electrical charge, dubbed "electrons." By measuring how cathode rays were deflected by electric fields, Thomson deduced the ratio of the electron's charge (denoted *e*) to its mass.

The idea that electrical charge is grainy seemed to verify American scientst Benjamin Franklin's intuition in the eighteenth century that electricity is not a smooth fluid but is rather composed of tiny particles. What, though, *is* the charge on an electron? Could we even be sure it was the same for every electron? Thomson set out to answer that by measuring *e* itself. He made use of a phenomenon discovered in the 1890s by the Scottish physicist Charles Wilson, whereby energetic particles passing through a chamber saturated with water vapor may induce ionization of molecules, giving them a charge and triggering their aggregation into water droplets. The idea was that one used a radioactive source to create ions and stimulate the formation of a tiny cloud inside a chamber. Then *e* might be measured by dividing the total charge on the cloud by the number of droplets it contained, assuming that each droplet had only one unit of the electron charge. The number of droplets could be estimated from their size (deduced from how quickly the cloud sinks down because of gravity, according to well-known principles of fluid flow) and the total volume of the cloud. But this method was fraught with assumptions and approximations. For one thing, it was hard to gauge the top of the little mist cloud, and the droplets evaporated rather quickly.

The American scientist Robert Millikan at the University of Chicago decided in 1907 that he could beat Thomson to the answer. He modified Wilson's method: instead of watching the cloud of water droplets sink, he pulled it upward using an electric field. If the droplets picked up a negative charge from electrons in the air ionized by radioactivity or X-rays, they would rise toward a positively charged plate above them.

Indeed they did—but so fast that they were almost instantly swept away. However, Millikan noted that a few lone droplets remained in the chamber. Presumably, these had just the right ratio of charge to mass for the upward pull of the electric field almost perfectly to balance the downward tug of gravity. Millikan realized that instead of trying to make difficult measurements of the motion of an entire cloud, he could literally focus instead on single droplets. These were typically just a few thousandths of a millimeter across, but could be seen using a microscope lens in the viewing port of the cloud chamber.

Millikan created an apparatus in which charged water droplets from a mist fell down through a small hole into the chamber, where their position could be tracked in the crosshairs of a microscope. At first the droplets would be left to fall under gravity; from their speed, their diameter could be calculated. (The diameter couldn't simply be measured using the microscope because the drops were so small that

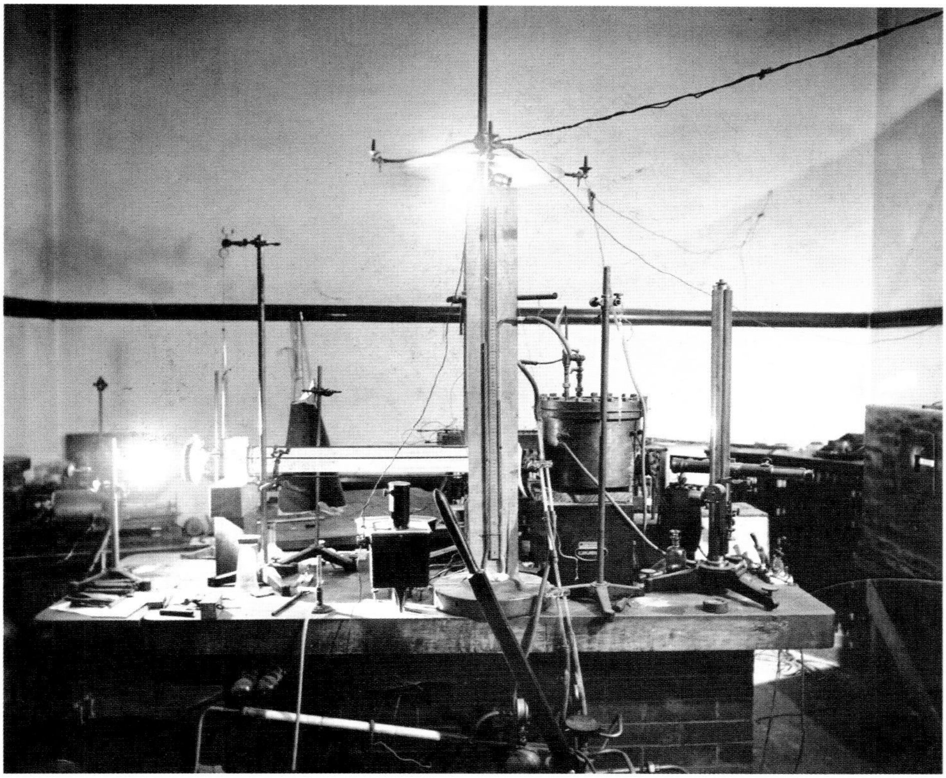

Robert Millikan's apparatus for his oil-drop experiment to measure the electrical charge of a single electron, 1909, courtesy of the Caltech Archives, California Institute of Technology, Pasadena.

light interference prevented their edges from being pinpointed.) Then the electrical field was applied, and the charge on a droplet could be deduced from its rate of ascent.

Millikan published his first results from this "balanced trap" method in 1909, but they were inconclusive because the water evaporated fast. He set his student Harvey Fletcher the task of coming up with a better approach, and Fletcher decided to replace the water with a nonvolatile oil that was used to lubricate clocks. From a drugstore he bought an atomizer of the kind used to spray perfumes, to create a mist of fine oil droplets that fell into the viewing chamber.

With painstaking practice Millikan and Fletcher became adept at manipulating droplets at will by adjusting the field strength, and at coping with their sensitivity to the slightest air currents and the random buffeting—Brownian motion (see page 104)—of the surrounding air molecules. They "executed the most fascinating dance," Fletcher said.

Millikan published his first oil-drop measurements in 1910, reporting that the differences between their electrical charges always seemed to be close to multiples of some minimal value. (Millikan's papers carried only his name—Fletcher was somewhat chagrinned, with good reason perhaps, at being left off.) This minimum value, he concluded, was the fundamental grain of charge: the charge on a single electron. Refining his experiment over the next three years, in 1913 he announced that e had the value 4.774×10^{-10} electrostatic units, the standard unit of charge at that time. That's an extraordinarily tiny value, and it was a tremendous challenge to show that the droplets' charges really did change in these minuscule discrete steps rather than smoothly, given that each measurement had many sources of potential error. Millikan was careful to make several observing runs for each droplet, so that he could take averages to reduce these uncertainties. All the same, his findings were disputed by the

ROBERT MILLIKAN
1868–1953

Having honed his skills as an experimental physicist who could measure small quantities with great accuracy, Robert Millikan followed up his work on the electron charge by verifying Albert Einstein's predictions of the photoelectric effect, a key empirical foundation of quantum theory. In doing so, he obtained a measure of another fundamental constant: Planck's constant, the basic "grain size" of the quantum world. His later work clarified the nature of cosmic rays from space.

See also: Experiment 27, The cloud chamber, 1894–1911 (page 121); Experiment 36, The photoelectric effect, 1899–1902 (page 155).

physicist Felix Ehrenfast in Vienna, who claimed that his own experiments revealed smaller charge steps—"subelectrons," as it were.

Ehrenfast's claims were soon discounted, however, and Millikan's were celebrated as disclosing one of nature's "fundamental constants"—indeed, as Millikan himself asserted (arguably with a little exaggeration), "probably the most fundamental and invariable entity in the universe." He was awarded the 1923 Nobel prize in physics for the work.

It would seem that challenges like Ehrenfast's may have made Millikan rather defensive, and it appears he was not entirely honest about what he had done. Although in his 1913 paper he wrote that the measurements he reported were "not a selected group of drops but represents all of the drops experimented upon during 60 consecutive days," later examination of his laboratory notebooks revealed this to be untrue. Rather, Millikan discarded some of his measurements.

Some have accused Millikan of picking only those results that "fitted" his conclusion about the value of *e* and throwing out ones that did not. But it seems instead that he was following his intuition regarding which readings were reliable. His notes are filled with comments like "Beauty" or "Perfect" for some runs, or, alternatively, "Something wrong" for others. These do not refer to the calculated values of *e* but to the raw observations: they reflect Millikan's feeling for when the experiment had worked well or succumbed to some disturbing influence such as an air current. It seems that Ehrenfast, in contrast, used all his measurements regardless of any perception of their quality.

It may feel like a dangerous, even unethical way to conduct an experiment. But, in fact, it reflects the reality of much scientific experimentation.

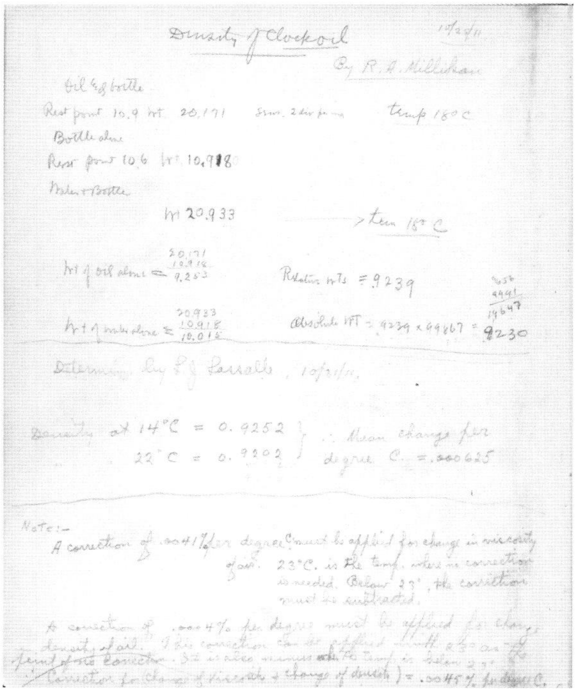

"Diversity of Clock Oil." From Robert Millikan's Oil Drop Experiment Notebooks, Notebook 1, October 1911 to March 1912, courtesy of the Caltech Archives, California Institute of Technology, Pasadena.

Robert Millikan's oil-drop experiment apparatus (reconstructed in 1969), courtesy of the Caltech Archives, California Institute of Technology, Pasadena.

If you are working at the edge of what can be detected or measured, or with a highly complex system, there's ample scope for the process not to work quite as it should. Experimentalists develop a feel for their equipment, getting to know its quirks and foibles. This is why, for example, some experiments are hard to replicate without the guidance of those who first conducted them, and why students have to spend some time getting to know the apparatus before it does what it should. Millikan's viewing microscope disclosed a theater in which each droplet was an actor with its own personality, and he needed good judgment to assess each performance.

Where Millikan surely did err is in not being truly honest about this: in not explaining that his data had been selected from a larger set of measurements. Experimenters are expected to be upfront about their methods and their choices. Even today, many experimental papers showing "representative" results such as microscope images are, in fact, often probably showing the best of these. There is no easy answer, no rigid set of criteria for reporting your results—for the fact remains that experimental science is not always an exact science, but must always to some degree be an art, requiring judgment and intuition.

Invention of the scanning tunneling microscope and moving single atoms (1981-1982/1989)

Q Can we see and move individual atoms?

In December 1959, physicist Richard Feynman gave a talk at the California Institute of Technology titled "There's plenty of room at the bottom." He said: "What I want to talk about is the problem of manipulating and controlling things on a small scale." By small, Feynman included the smallest conceivable: "What would happen if we could arrange the atoms one by one the way we want them?" That feat became possible sooner than Feynman imagined, thanks to experiments conducted just over two decades later.

In the mid-1950s, researchers at Pennsylvania State University reported the first images of individual atoms, obtained with a device called the field ion microscope. It used high voltages to pull electrons off gas atoms stuck to the surface of a metal needle. The ionized atoms were then repelled from the tip and pushed toward a phosphor screen, projecting a kind of enlarged image of the tip itself in which the atoms showed up as bright dots.

Field ion microscopy only offers atomic-resolution images for special kinds of sample, though. In the early 1980s, the physicists Heinrich Rohrer and Gerd Binnig at IBM's research laboratories in Rüschlikon, near Zurich in Switzerland, invented a new kind of microscope that could provide images of individual atoms with much greater versatility.

IBM was one of several industrial tech giants operating research labs where some employees were free to pursue whatever science they liked, conducting open-ended research every bit as creative and speculative as that at academic institutions. In the late 1970s, Rohrer was trying to understand a phenomenon called quantum tunneling, but he lacked the tool he needed, so he set out to invent it. In quantum tunneling, the wave-like nature of tiny objects described by quantum mechanics, such as electrons, allows them to tunnel through a barrier even though classically they have insufficient energy to do so. Rohrer allowed electrons to tunnel between a charged, fine metal tip and a surface of some kind, holding the tip very close to the surface so that electrons could jump between them. He suspected such a device could be used for spectroscopy—measuring the energy states of atoms and materials—on very small samples.

After Rohrer recruited Binnig on the project, the two realized that this setup might also act as an imaging device. If the tip was held just a nanometer (a few millionths of a millimeter) or so above the sample surface, the rate of electron

HEINRICH ROHRER | 1933–2013

The Swiss physicist Heinrich Rohrer was no stranger to sensitive experiments when he devised the scanning tunneling microscope. For his work on superconductivity at Zürich in the 1950s he also had to work at night so that his experiments were not disturbed by road vibrations. He joined IBM in 1963, where such work on exotic aspects of physics was then enthusiastically welcomed by the information-technology company, who figured—rightly—that it could ultimately have technological spinoffs. In later life, Rohrer became a prominent advocate for nanotechnology: science at the scale of nanometers.

Scanning tunneling microscope-MHS 2237, 1986, *Musée d'histoire des sciences de la Ville de Genève*, Lake Geneva, Switzerland.

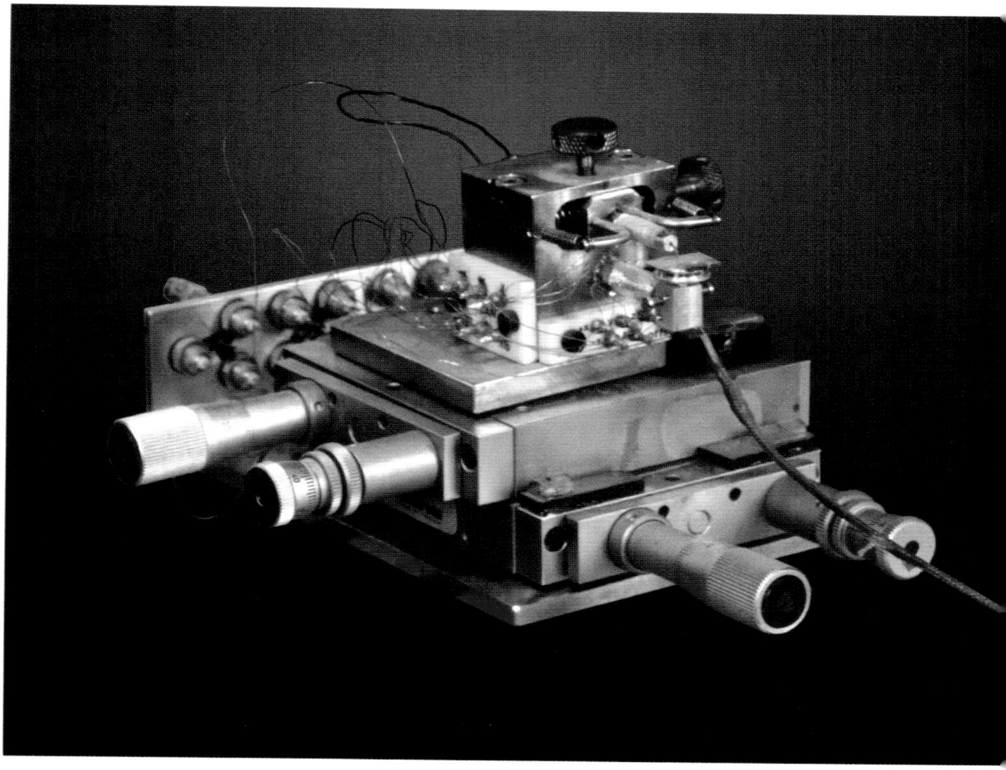

tunneling should be very sensitive to the distance separating them. As the tip was scanned over the surface, any tiny bump that decreased the gap should substantially boost the electric current in the tip produced by tunneling. So, from the strength of this current, one might map out the topography of the surface by scanning along transects. Perhaps that map might reveal the arrangement of the surface atoms themselves, each atom appearing as a bright bump, where they sit side by side like eggs in an eggbox.

Rohrer later averred that it was probably his and Binnig's very lack of experience in microscopy or surface science that "gave us the courage and light-heartedness to start something which should 'not have worked in principle,' as we were so often told." While it's true that some experiments demand tremendous expertise in a field, others may rely on an almost naïve recklessness in the

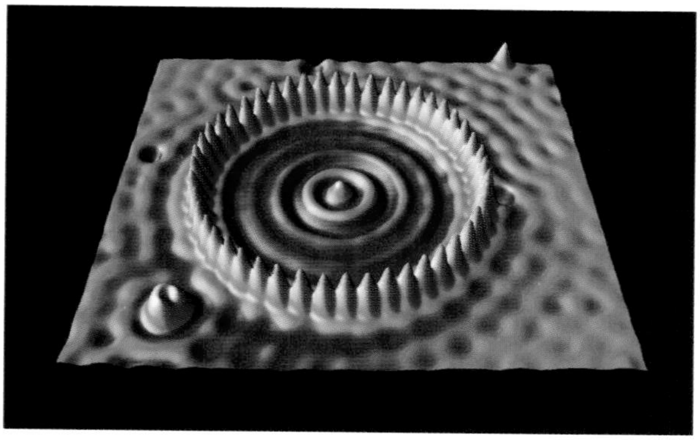

A micrograph taken by a scanning tunneling microscope of a quantum corral. This artificial structure is created from forty-eight iron atoms (with red peaks) arranged on a copper surface, IBM Almaden Research Center, San Jose, California.

GERD BINNIG | B. 1947

Born in Frankfurt, Gerd Binnig studied physics in his home city before joining IBM in 1978 to work with Swiss physicist Heinrich Rohrer. He has stayed with the company for the rest of his career, working in its division in San Jose before founding a new physics group in Munich. He has also founded a company that works on computer image analysis. The Nanotechnology Center at IBM Rüschlikon is now named after Binnig and Rohrer.

experimenters to attempt what "experts" have deemed impossible.

Could a metal tip really be made sharp enough to distinguish atoms? Surely even the finest needle tip would be hopelessly blunt at the atomic scale? But Rohrer and Binnig realized that their tips could actually be self-sharpening: the intense electric field at the extremity would pull and evaporate atoms until the tip developed a pyramid-like shape that narrowed almost to a single atom. To position this tip precisely, the researchers used piezoelectric materials, which deform in response to electric fields, to make a scanning system with no moving parts.

The biggest problem was keeping the tip steady just a nanometer or so above the surface. The slightest vibration might send it crashing into the sample like a dropped record stylus. Binnig and Rohrer, assisted by their technicians Christoph Gerber and Edi Weibel, made rubber shock-absorbers, or suspended their prototype devices from rubber bands, or used superconducting supports levitated in a magnetic field. They measured at night when the labs were empty and the roads quiet, and, Rohrer recalled, "hardly daring to breathe"—to avoid disturbances, but also from excitement.

The researchers reported in late 1981 that they could measure a tunneling current between a platinum surface and a tungsten tip. The following year they made measurements on gold and silicon surfaces, and to their astonishment and delight saw that the scanned maps showed arrays of bumps corresponding to the individual atoms. "I couldn't stop looking at the images," said Binnig. "It was [like] entering a new world."

The IBM team called their device a scanning tunneling microscope (STM). By 1983 they were turning their attention to biological molecules, scanning the outline of a strand of DNA lying on a surface. The STM was not especially hard to make, and very soon researchers all over the world were building them. One of the limitations, however, was that the objects they imaged had to be somewhat electrically conducting, so that electrons could tunnel into or out of them. But many samples of interest, especially most biomolecules, were insulating. Binnig, working with Gerber and Calvin Quate of Stanford University, found a way around that problem by

DON EIGLER | B. 1953

The images obtained by Don Eigler and his coworkers in the 1990s using the STM have become some of the most iconic emblems of nanotechnology. They show atoms arranged using the instrument's needle tip and ripples in the underlying surface due to the quantum nature of the mobile electrons it contains, all artfully colored as if they are futuristic virtual landscapes. In such ways, Eigler, working at IBM Almaden near San Jose in California, has combined his experimental imagination with eye-catching visual flair. Eigler envisaged manipulating individual atoms and molecules to conduct chemical reactions, a prospect now realized at IBM's Rüschlikon laboratories.

See also: Experiment 23, Understanding Brownian motion, 1908 (page 104); Experiment 37, The diffraction of X-rays by crystals, 1912 (page 158).

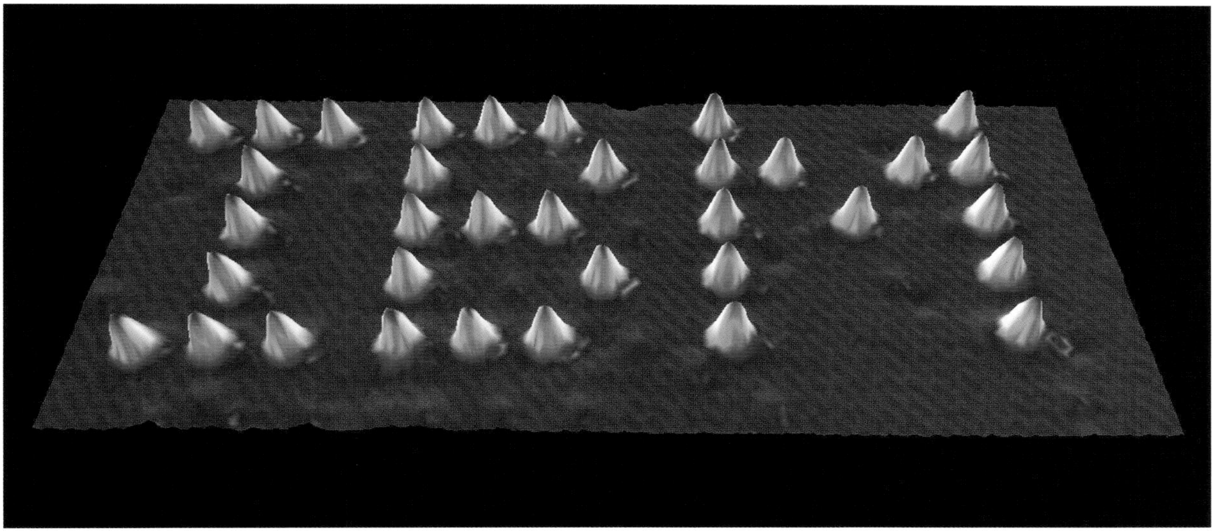

"IBM" spelled out using xenon atoms on a nickel substrate, demonstrating the capacity of a scanning tunneling microscope to manipulate matter at the level of individual atoms, April 1990, IBM Almaden Research Center, San Jose, California.

making a similar scanning device in which the needle tip could be pulled more or less close to the surface by the force of attraction between them. Such an attraction, called the van der Waals force, exists for all substances, and also depends on the separation between the two objects. By using a feedback circuit that keeps the sample-tip separation constant, the researchers could measure how the force changed as the tip was scanned. This force map, like the map of tunneling current in the STM, acts as a proxy for the surface topography.

Binnig and colleagues unveiled this "atomic force microscope" in 1986—the same year in which he and Rohrer shared the Nobel prize in physics for their invention of the STM. Such a short time from discovery to Nobel is unusual, and testament to the immediate impact the work had across many areas of science.

The forces between tip and sample are present in the STM too—and that can be a problem. When Don Eigler, working at IBM's laboratory in Almaden, near San Jose in California, was using the device at the start of the 1990s, he found to his frustration that sometimes "the atoms would not hold still," owing to the influence of the tip. But Eigler figured that "I can get that under control and use that [force] to put atoms where I want them to go." He tried it out with a single atom of the inert gas xenon resting on the surface of platinum. By carefully changing the voltage of the tip, Eigler found he could drag the xenon atom to a new, chosen location and then let it go. Scanning probe microscopes are not just probes for studying the atomic world but tools for intervening in it.

In 1990 Eigler, with his colleague Erhard Schweizer visiting from Berlin, marshaled xenon atoms on a nickel surface with the STM tip until they spelled out, as if in a dot-matrix printer, the company's logo: "IBM." As Eigler said: "We got to Feynman's 'all the way down.'" That iconic image, spelled out in thirty-five atoms, has become probably the most famous image in the history of nanotechnology, the discipline that emerged from Feynman's prophetic vision.

The particle zoo

Atoms are not nature's finest grain—they are not, after all, indivisible. That atoms have a *subatomic* structure was one of the astonishing revelations of the early twentieth century, and figuring out the nature of the constituent particles and the rules by which they are combined became one of the major projects of the physical sciences. Disconcertingly, that endeavor seemed not so much to simplify as to complicate our view of matter—for the particles that make up atoms turned out to have cousins that play no role in the ordinary stuff of the world. As one scientist exclaimed in bafflement when the first of these came to light: "Who ordered that?" Having finally made sense of the profusion of chemical elements, now the task for these scientists was to bring order to the growing menagerie of the subatomic world. It seems we have finally "completed the set," as it were—but this has not answered all the questions, nor stilled suspicions that there are yet more fundamental particles to be discovered.

The cloud chamber (1894-1911)

Q How can we detect particles that are too small to see?

When you open a new window on the world, you never know what you're going to see. Scientific instruments designed for one purpose might turn out to be just what is needed for another application entirely. Or they might bring to light a phenomenon no one anticipated: an unguessed aspect of reality.

Both apply to the invention of Charles Thomas Rees Wilson (universally known by his initials C. T. R.), who wanted to make clouds. Having studied the physical sciences at Cambridge University, he began working in its Cavendish Laboratory in the mid-1890s. On a visit to his homeland of Scotland in 1894, he scaled the highest mountain, Ben Nevis, during a rare spell of good weather and was able to see the optical effects called glories and coronas created as sunlight fell onto the clouds and mist below the peak. He decided to explore the physics of these phenomena in the laboratory.

Wilson devised an apparatus to make "artificial clouds" in an enclosed chamber. Clouds form when water vapor in the air condenses into tiny droplets. Wilson triggered droplet formation in his chamber (a glass jar) by allowing moist air flowing into it to expand, dropping the pressure—basically the same process that produces a wisp of mist at the mouth of a bottle of cold sparkling water when it is opened and the internal pressure released.

But droplet formation generally requires the water to have something to condense *onto*. Typically, that happens in the atmosphere on small particles called aerosols, such as dust or tiny salt particles from the oceans: these act as so-called condensation nuclei. Wilson took care to filter any dust out of his air flow, however. So what was seeding the formation of droplets? He figured that these condensation nuclei must be tiny, and speculated that they were electrically charged molecules (ions) present in the air.

In 1895, Wilhelm Röntgen discovered X-rays, a kind of radiation that was able to cause ionization of air. Early in the following year, Wilson found that exposing his "cloud chamber" to X-rays led to denser clouds, supporting his idea that ions act as the condensation nuclei. From then until 1900, he experimented with his cloud chamber while employed by the UK's Meteorological Council to study atmospheric electricity.

Like many experimental scientists of his time, Wilson couldn't get his equipment off the shelf, but had to make it himself. This meant blowing glass into the flasks, bulbs, and coils he needed, and then joining the glass parts together by

C. T. R. WILSON | 1869–1959

Charles Thomas Rees Wilson was born into a family of Scottish farmers and studied at Owen's College, which later became Manchester University. Despite the impact his cloud chamber had on nuclear physics, he remained essentially a meteorologist and made significant contributions to our understanding of thunderstorms.

See also: Experiment 28, Discovery of the positron, 1932 (page 124); Experiment 29, Detection of the neutrino, 1956 (page 128).

C. T. R. Wilson's experiments with the cloud chamber resulted in a new technique for discovering fundamental particles. His original 1911 cloud chamber, shown here, is kept at the Cavendish Laboratory, University of Cambridge.

grinding them to just the right shape to make an airtight seal. The first cloud chambers didn't look like much, but it took great skill and patience to make them, and there were no doubt many frustrating breakages.

These instruments were essentially detectors of ionizing radiation, which included the alpha and beta particles emitted by radioactive elements. They soon became an indispensable tool for the nascent field of nuclear science—even though that was far from Wilson's initial interest in atmospheric science.

In 1911, Wilson showed that he could see the tracks left by single particles moving through the chamber, leaving a trail of ions (and thus droplets) in their wake, rather as high-altitude aircraft leave vapor trails in a clear sky. Wilson

Alpha particle tracks observed in C. T. R. Wilson's 1912 cloud chamber experiments. These images were among the first Wilson obtained using his perfected cloud chamber for photographing particle tracks. From Wilson's *Philosophical Transactions of the Royal Society*, A87, 277, 1912, the Royal Society, London.

would shine a light on the tracks to make them more apparent and then photograph them: these wispy threads showed the trajectory of the invisible particles. Those tracks could reveal the particles' properties: when the chamber was placed between the poles of a magnet, they would curve and spiral, in one direction for positively charged particles, and the other for negative ones. How tight the curve is depends on the charge, mass, and speed of the particle, and so cloud chambers became a tool for investigating the fundamental properties of subatomic particles, their value testified by the award of the 1927 Nobel prize in physics to Wilson.

Discovery of the positron (1932)

Q What are high-energy cosmic rays?

Before C. T. R. Wilson's cloud chamber (see page 121), the main instrument for studying ionizing radiation was the electroscope, a simple device invented by Pierre and Marie Curie. Two hanging sheets of gold leaf were electrically charged so that they repelled one another and made an inverted V-shape. If the air between them was ionized, their charge leaked away and they dropped back toward one another. Using such an instrument, in 1911 and 1912 the Austrian physicist Victor Hess ascended to more than 17,400 feet altitude in a hot-air balloon and measured the background radiation levels of the atmosphere. He discovered that it was being flooded with ionizing rays coming from space, dubbed cosmic rays, whose source was unknown.

In the late 1920s, Robert Millikan at the California Institute of Technology studied cosmic rays using a cloud chamber, joined by his PhD student Carl Anderson, who gained his doctorate in 1930. Anderson used a modified cloud chamber in which alcohol vapor rather than water vapor condensed to droplets, because it produced brighter trails. To bend the tracks and deduce the charge of the particles generating them, Anderson built a water-cooled electromagnet that consumed so much power that he had to run his experiments at night so as not to starve the rest of the laboratory of electricity.

Anderson set out to investigate the claim made in 1929 by Russian scientist Dmitri Skobeltsyn that he had seen cosmic-ray tracks that hardly bent at all, meaning the particles must have a great deal of energy. Were they real? If so, what were they?

In 1932, Anderson obtained more than a thousand photographic images of particle tracks and combed through them for anything interesting or unusual. This is reminiscent of the way today physicists working with particle colliders have to sift through immense amounts of data collected by their detectors to seek signs of the kinds of particles or interactions they have predicted. But today that process is automated by computer; Anderson had to do it by hand.

Among his pile of photographic plates, he spotted fifteen trails that were bent by the magnet in a way that indicated the particles responsible had a positive charge. However, they only bent a little, suggesting that they were very light—much more so than the lightest positive particle then known, the proton that appeared in atomic nuclei.

CARL ANDERSON | 1905–1991

Carl Anderson came from a Swedish family who emigrated to New York, where he was born. He graduated from the California Institute of Technology and pursued his doctoral thesis there under Robert Millikan. After discovering the positron, he and his student Seth Neddermeyer found another new fundamental particle, the muon (a kind of heavy electron), in 1936.

Carl Anderson with the cloud chamber, with which he discovered the positron (anti-electron) in 1932, so proving the existence of antimatter, California Institute of Technology, Pasadena.

PAUL DIRAC | 1902–1984

Paul Adrien Maurice Dirac, born in Bristol, England, is considered one of the most significant theoretical physicists of his generation. He was renowned for being precise and sparing with his words: his Cambridge colleagues invented the unit of speech rate of a "dirac"; one word per hour. Dirac reformulated Erwin Schrödinger's quantum wave mechanics in a way that was consistent with special relativity. For their work on quantum mechanics, Dirac shared the 1933 Nobel prize in physics with Schrödinger.

See also: Experiment 27, The cloud chamber, 1894–1911 (page 121); Experiment 29, Detection of the neutrino, 1956 (page 128).

These new particles seemed to have a mass and charge about equal to that of the electron, but of opposite sign. At first, Anderson is said to have suspected that these were just ordinary electrons, and that some prankster had reversed the polarity of his magnetic field. Anderson called the new particles "positrons"—or rather, he accepted that name at the suggestion of the editor of the journal in which he published the paper announcing his discovery.

Could such entities exist, though? In one of those curious coincidences that abound in scientific discovery, something like a positive electron had been predicted theoretically just two years earlier by the British physicist Paul Dirac. By reformulating the theory of quantum mechanics, which described the behavior of very small particles like atoms and their constituents, to bring it in line with special relativity, Dirac found that his ideas required the existence of electron-like particles with negative energy—an apparently nonsensical idea which he interpreted as electrons that behave as though they have a positive charge.

After the mathematical physicist Herman Weyl showed in 1931 that this "negative-energy electron" must have the same mass as an ordinary electron, Dirac called it an antielectron and showed that if it met a normal electron, the two would annihilate one another with the release of energy.

In essence, Dirac had predicted antimatter. Such a bold proposal was too much for many of his contemporaries, who dismissed the idea. But the antimatter partner of the electron was expected to have exactly the properties Anderson observed in his positron. Once his cloud-chamber images became known, others tested the idea. In particular, physicists Patrick Blackett and Giuseppe Occhialini, working with cloud chambers in the Cavendish Laboratory, in Cambridge, quickly confirmed that positrons were real and that they were created along with electrons from a high-energy cosmic gamma ray, which could spontaneously generate the particle pairs by the conversion of energy to matter. Anderson had, then, discovered not just a new particle but the existence of Dirac's antimatter. In 1936, aged thirty-one, he became the youngest person ever to receive the physics Nobel prize, which he shared with Viktor Hess.

Every known particle is now known to have an antimatter twin. One of the deepest unsolved mysteries in physics is why our universe seems to have more matter than antimatter, since theories of the Big Bang predict that they should have been created in equal quantities.

Carl Anderson's cloud chamber photograph showing the tracks of a positron, August, 2 1932, California Institute of Technology, Pasadena.

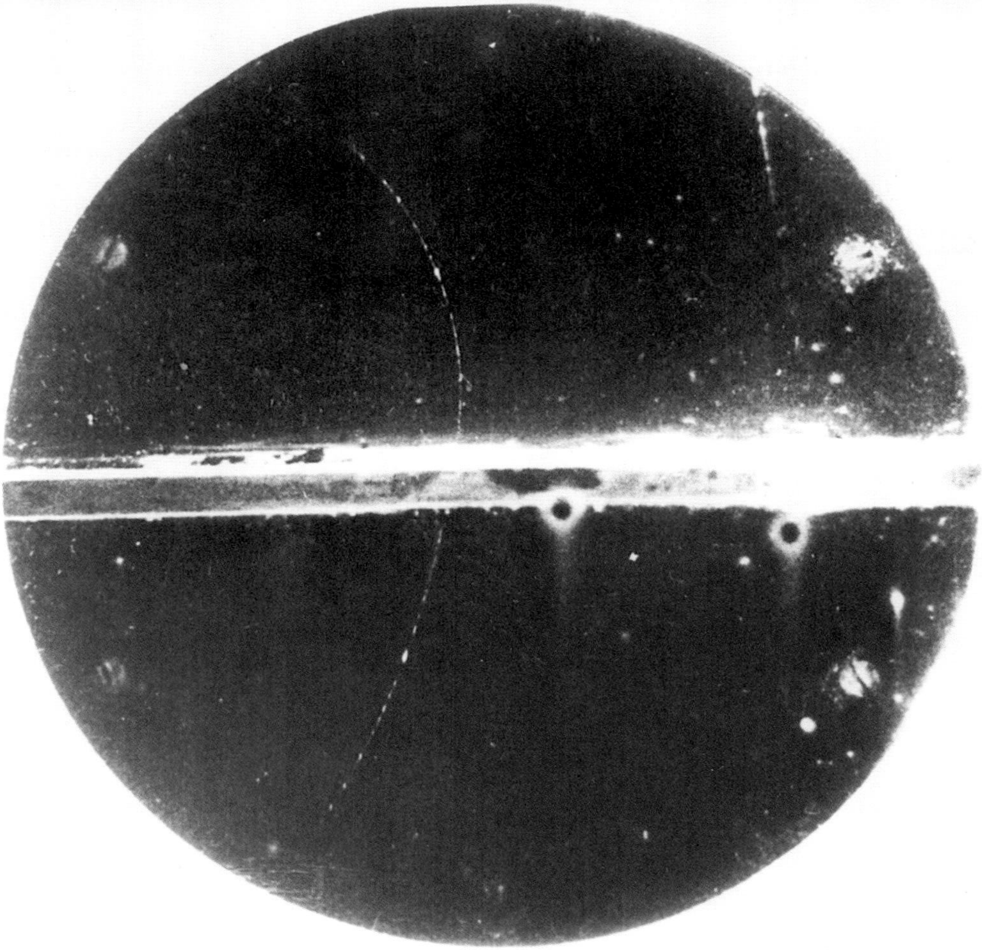

Detection of the neutrino (1956)

Q If the hypothetical particle called the neutrino barely interacts with anything, how can it be detected?

"I have done a terrible thing," the Austrian physicist Wolfgang Pauli confessed in 1930. "I have postulated a particle which cannot be detected." It was terrible because it seemed unscientific: you are not meant to formulate a hypothesis that can't be tested (and so disproved).

Why would Pauli commit such a solecism? It was an act of desperation compelled by the puzzle of radioactive decay by emission of a beta particle. This particle is nothing but the familiar electron—except that, weirdly, it comes from inside the nucleus of the decaying atom, where scientists figured that only protons and neutrons reside. In effect, beta decay happens when a neutron, with no electric charge, transforms into a proton, with a positive charge, and an electron, with an equal but opposite negative charge. The phenomenon was perplexing because the electron emitted could have a range of energies, seemingly at random, up to some maximum value. But the principle of conservation of momentum seemed to demand that its energy should be the same each time.

Pauli suggested that maybe the balance of energy was carried off by another particle, so that their combined energy was always fixed. If so, the particle must have no charge and no mass—which made it virtually impossible to detect. He called it a neutrino. Soon after Pauli proposed it, other physicists estimated that a particle of this sort should be able to pass through a lead barrier several light years thick. To detect such an entity looked like a hopeless task.

By the early 1950s many nuclear physicists felt compelled to accept that Pauli was probably right. Yet they couldn't be sure until someone actually saw a neutrino. US physicist Fred Reines, working at the Los Alamos National Laboratory, which arose from the Manhattan Project, decided to try.

Sure, any given neutrino might be impossibly elusive. But quantum physics implied that interactions between particles are determined by probabilities. So long as the chance of a neutrino colliding with another particle was not exactly zero, such an event might happen, however rarely. Given enough neutrinos passing through a detector, you might just get lucky.

Reines imagined making a detector from a tank of liquid, in which such a collision with a molecule in the liquid might produce a brief flash of light that could be picked up by photosensors lining the tank. But the volume of such a detector for neutrinos, and of the liquid it would contain, would have to be enormous. Using estimates of how frequent such collisions might be, Reines did a back-of-the-envelope calculation and discovered

FREDERICK REINES | 1918–1998

Born in New Jersey, Frederick Reines studied engineering at the Stevens Institute of Technology in Hoboken, New Jersey, before being recruited by Richard Feynman to work on the Manhattan Project to develop the atomic bomb. A delegate at the Atoms for Peace conference in Geneva in 1958, he later participated in searches for the neutrinos produced by supernovae (exploding stars). He won the Nobel prize for physics in 1995.

Project Poltergeist: Clyde Cowan (far left) and Frederick Reines (far right) with the Poltergeist team, 1955, Los Alamos National Laboratory, New Mexico.

that to have much chance of spotting a neutrino this way he would require a detector thousands of times larger than any previously made.

After Reines happened to meet and get talking to chemical engineer Clyde Cowan at an airport, the two decided to give it a try anyway, and in 1951 they launched what they jokingly called Project Poltergeist: a search for something incredibly elusive. In a warehouse at Los Alamos they assembled a double tank of liquid (106 US gallons in all; the final detector used a solution of cadmium chloride) surrounded by ninety devices called photomultiplier tubes, which amplified faint flashes into electrical pulses that would be recorded as blips on an oscilloscope. According to theory, when a neutrino collides with a proton, it should produce two flashes five-millionths of a second apart: a distinctive signature. But Reines and Cowan had to eliminate false signals from other sources, such as cosmic rays or nuclear decay processes, which they did by surrounding the tank with shielding made from blocks of paraffin wax.

Once the detector was assembled in 1953, they took it to a nuclear reactor site at Hanford in Washington, because nuclear decay processes in the reactor were predicted to generate abundant neutrinos. But the background signal from cosmic rays proved too big to see anything conclusive. So the duo then moved their detector all the way to South Carolina, where another reactor was located at Savannah River. Here they could set up the device in a basement 39 feet below the reactor, where it would be sheltered from most cosmic rays. In 1955, they began to see what looked like neutrino signals. Crucially, there were five times as many of these when the reactor was operating than when it was switched off. In June they felt able to telegram Pauli in Zürich, saying: "We are happy to inform you that we have definitely detected neutrinos from [nuclear] fission fragments."

CLYDE COWAN | 1919–1974

Graduating with a degree in chemical engineering in 1940, Clyde Cowan worked for the US Air Force on defence against chemical weapons. He joined Los Alamos in 1949, teaming up with Reines two years later. He would surely have shared the Nobel prize if he had not died before the award was made.

See also: Experiment 27, The cloud chamber, 1894–1911 (page 121); Experiment 28, Discovery of the positron, 1932 (page 124).

30

Discovery of the Higgs boson (2012)

Q The Higgs particle is supposed to give mass to other particles—but is it real?

The Large Hadron Collider (LHC), the particle accelerator at the European center for particle physics CERN, near Geneva in Switzerland, has been called the biggest machine on Earth. It is indeed hard to think of any device to compare in scale with the 17-mile-long tunnel in which protons are accelerated by gigantic, powerful magnets until they have velocities 99.999999 percent of that of light, before they are sent crashing into each other in the hope that among the debris will be new fundamental particles.

The LHC could hardly be further from the traditional notion of an experiment as something done by a scientist at a laboratory bench. The machine took fourteen years to build and test before, in 2008, it began colliding protons; it cost in the region of $4.75 billion; it produces so much data that we could never analyze the results without the help of dedicated supercomputers; and it is used and maintained by an international army of thousands of scientists and technicians. All this is for the purpose of understanding how the universe exists at all.

Particle accelerators were conceived as instruments for probing inside atoms. Ernest Rutherford's discovery of the atomic nucleus used alpha particles emitted by radioactive decay as projectiles that could penetrate into the atom and disclose its internal structure. But there was no control over the properties of such projectile particles—their energy and mass, for example. In the late 1920s, Rutherford, while working at the Cavendish Laboratory in Cambridge, longed for more energetic projectiles that could penetrate deeper. A doctoral student named Ernest Walton suggested a solution: to use electromagnetic fields to accelerate charged particles to higher energies. Between 1927 and 1930, Walton and John Cockcroft, another researcher based at the Cavendish, put together the first particle accelerator, using enormous voltages to accelerate protons. In 1932, they fired these particles at lithium atoms, inducing them to decay radioactively: they had managed to "split the atom" by artificial means.

To reach even higher energies, American physicist Ernest Lawrence at the University of California at Berkeley had the idea of bending the particle trajectories into an expanding spiral using magnets, giving them a fresh push on each circuit. He demonstrated the first of these "cyclotrons" in the early 1930s, and he and others were soon using them to bombard and transmute other atoms. In the late 1940s, scientists in the USA began to devise accelerators that held the particle paths in a circle, rather than an expanding spiral. This was more complicated because it meant that, as the particle energy was increased by acceleration, the confining magnetic field had to be continually adjusted to keep the particles on the same circular orbit. Such machines were called synchrotrons.

Both circular and linear particle accelerators grew steadily in size, energy, and ambition over the following decades, becoming gradually more expensive to build and needing dedicated teams of engineers to keep them running. This was now "big science," done on an industrial scale. Construction of the Tevatron accelerator at the National Accelerator Laboratory (now known as Fermilab), near Chicago, began in 1969; with a budget of $250 million and a tunnel nearly 4 miles long, it switched on its proton beam in 1972.

A section of the Large Hadron Collider, CERN, near Geneva, Switzerland.

CERN was established in 1954 as the hub of experimental particle physics for member states of Europe and its environs. In 1983, construction began on the Large Electron-Positron Collider (LEP), which smashed those two particles together in a 17-mile circular tunnel about 100 feet underground on the Swiss–French border. The electrons and positrons would circulate in opposite directions before the beams were brought together, so that the particles crashed into one another and released a huge burst of energy. As with all colliders, the goal is that some of this energy is converted into mass: into new particles, perhaps including ones not seen before. While electrons, protons, and neutrons are all around us—we are made from them—many other fundamental particles don't survive for long before decaying. If we want to see them, we need to make them afresh. But because they typically have greater masses than protons and neutrons, it takes a lot of energy to produce them. Their fleeting existence may be registered by the

PETER HIGGS | B. 1929

Peter Higgs was awarded the 2013 Nobel prize in physics after the discovery of the particle named after him. After his three key papers were published in 1964, he not only made no further significant contributions to the field but barely published anything at all. He stayed at the University of Edinburgh for most of his career, but told reporters on the eve of his prize that he wouldn't be considered productive enough to get an academic job today.

FRANÇOIS ENGLERT | B. 1932

François Englert shared the 2013 Nobel prize with Peter Higgs; his mentor Robert Brout would surely have done so too had he not died in 2011. Born in Belgium to a Jewish family, Englert had to live in children's homes and hide his Jewish identity during the Nazi occupation. The theoretical framework developed by Englert and Brout, and independently by Higgs and others, laid the foundations for the theory that unifies the electromagnetic and weak nuclear forces as having once been a single "electroweak" force.

particle detectors installed at these colliders to analyze the debris of collisions. Typically, the chance of making such particles in any given collision is tiny, and so huge numbers of collisions must be induced, and their outcomes carefully sifted, to spot any sign of them.

That's the case for the Higgs boson. This particle was first predicted in 1964 by British physicist Peter Higgs, and at the same time by Robert Brout and his former research assistant François Englert at the Free University of Brussels, as well as by three other researchers. All proposed that "empty" space is pervaded by an energy field, now called the Higgs field, which keeps it stable. Some other particles "feel" the Higgs field and are slowed down by it as though by a kind of drag. This manifests as the particles acquiring mass. Without the Higgs field, fundamental particles like the proton and electron wouldn't have mass, and so there would be no gravitational attraction, no galaxies, stars, planets, or people. Since the particle of light, the photon, doesn't feel the Higgs field, it is massless.

In fundamental physics, all fields have particles associated with them, which might be regarded as a kind of local condensation or knot in the field. Just as the photon is the particle associated with electromagnetic fields, so the Higgs particle (a "boson" is a particle with certain quantum-mechanical features) is the particle linked to the Higgs field. Higgs and others predicted that it would have a lot of mass, and so would be hard to make—it would need very high-energy collisions. The LHC was constructed to produce these putative particles.

The collider was installed in the tunnels that previously housed LEP. Two detector systems—in effect, two separate experiments, using different methods—were built around the circumference to look for the Higgs boson: one is called ATLAS, the other CMS (the meanings of the acronyms are not very illuminating for nonspecialists). There are two other key experimental facilities attached to the LHC ring too—it was not all about the Higgs. But that was the main goal, for by the time the LHC began operating in 2008 this particle had become the last piece in the jigsaw created by physicists to describe all known particles and forces (except gravity), called the Standard Model. The discovery of the Higgs would complete the Standard Model and explain the origin of mass. That is, if the Higgs boson did indeed exist. This was by no means certain, although it was what the theory predicted. What's more, no one knew exactly what mass it would have, and thus what energy would be needed to make it. What if the

FABIOLA GIANOTTI | B. 1960

Italian physicist Fabiola Gianotti led the team on the ATLAS experiment at the Large Hadron Collider, one of the two that discovered evidence for the Higgs boson. In 2016, she was elected as the first female director general of CERN.

See also: Experiment 27, The cloud chamber, 1894–1911 (page 121); Experiment 28, Discovery of the positron, 1932 (page 124).

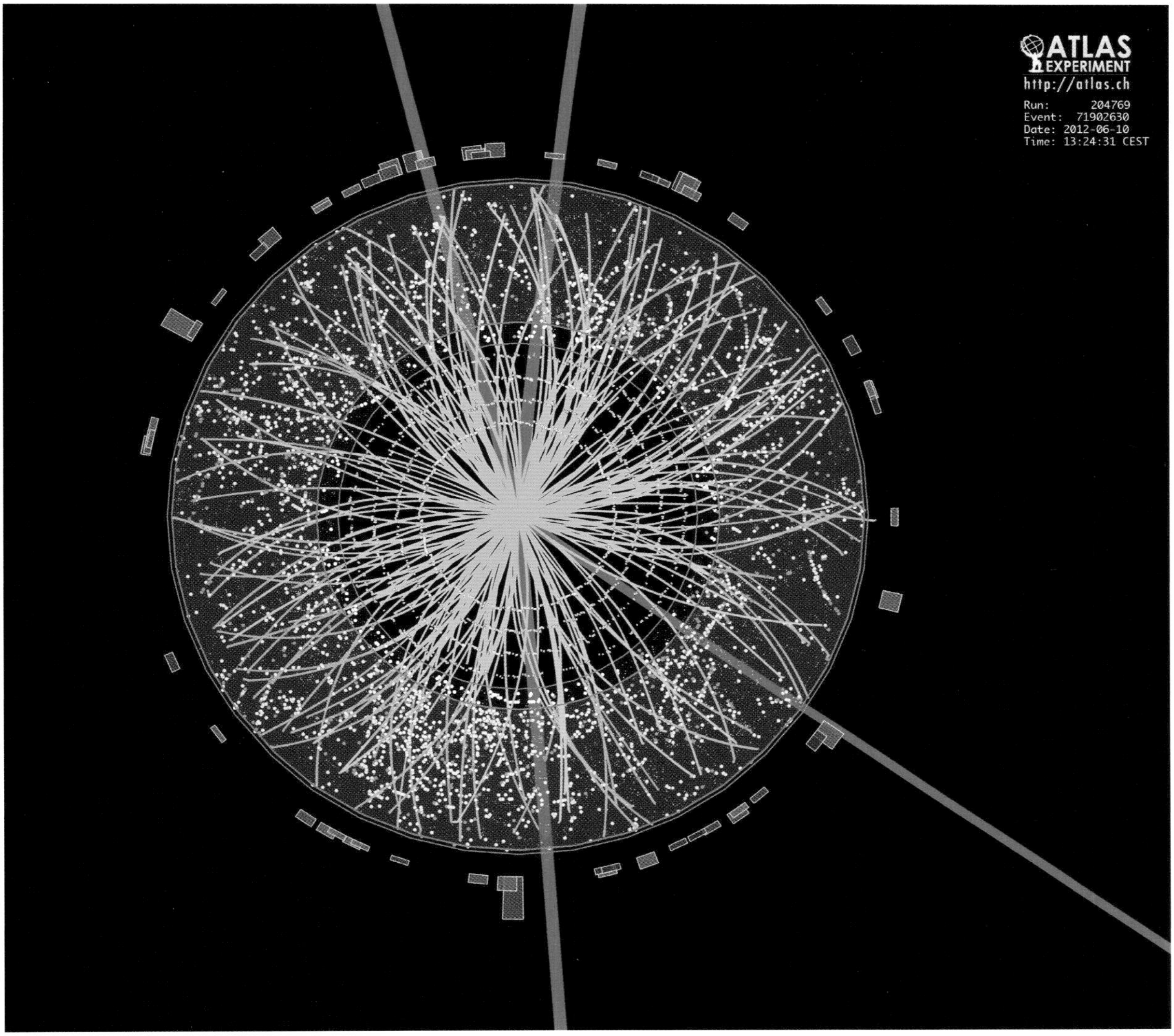

LHC was not powerful enough? But nine days after the switch-on, a flaw in one of the immense magnets used to keep the circulating proton beam in place led to an explosion that wrecked some of the tunnel. It took nine months before the machine was working again. Despite the setback, in July 2012 the CMS and ATLAS teams announced their findings. To cheers and even tears, especially those of the 82-year-old Higgs

Schematic transverse section through the ATLAS detector, showing the computer reconstruction of a proton-proton particle collision event, which was recorded on July 4, 2012 during the search for the Higgs boson.

who was in the audience, CERN's director general Rolf-Dieter Heuer told the world that the Higgs boson had been found.

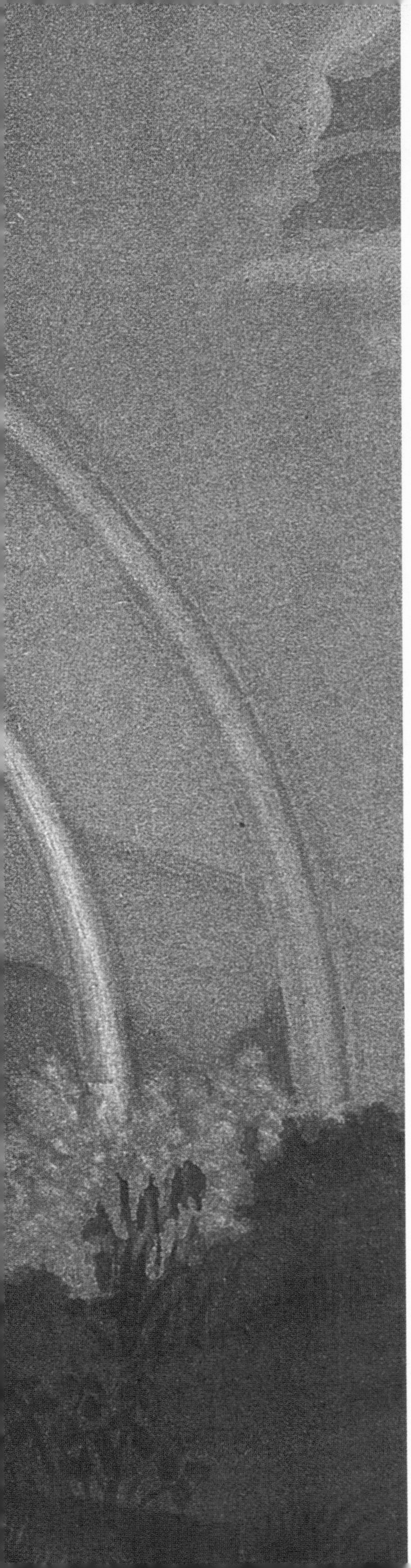

CHAPTER FOUR

What is light?

Light as waves: classical optics

Light is one of the oldest subjects of investigation, but it is also one of the most mysterious. It is central to our experience of the world, and yet it is immaterial, ephemeral, contingent. Light behaves in rule-bound ways that make it ideal for experimental study, and it commanded a central position in the natural philosophy of the Middle Ages and in early modern science. But we shouldn't forget that this fascination had theological roots: light was a symbol of the divine—never more so than when it decorated the sky with a rainbow. At the same time, there were practical benefits to understanding how to manipulate light: for making mirrors and eyeglass lenses, and later, in the use of heliographs for distant messaging. When experiments in the nineteenth century showed that light was somehow connected to both electricity and magnetism, it seemed clear that it could help reveal nature's deepest secrets.

31

The camera obscura (early 11th century)

Q How does a pinhole camera work?

Much of what we now know about the learning of ancient Greece, in particular about its science and philosophy, is due to the efforts of scholars in the Islamic world of the early Middle Ages. The lands under Islamic rule stretched from southern Spain to the Caspian Sea and modern-day Afghanistan and India. Here, Greek writers were translated into Arabic and studied with reverence, but their ideas were also challenged, amended, and extended. While it was once commonly thought that the Islamic world merely preserved the knowledge of Classical civilization, it's now clear that its scholars contributed much that was original, particularly in experimental science. The legacy of Islamic scholars can be seen today in several scientific words: algebra, algorithm, alkali, alcohol. In the Western Middle Ages and the Renaissance, some of these scholars were as revered as Greeks like Aristotle and Plato.

At the start of the Abbasid period in the eighth century, the caliph Al-Mansur transformed Baghdad into a citadel of learning and founded an institution called the House of Wisdom, an Islamic equivalent of the famous Library of Alexandria. Here scholars could learn about the astronomy of Ptolemy, the mathematics of Euclid, the physics and philosophy of Aristotle, and the medicine of Galen.

That era was waning by the time Abū ʿAlī al-Ḥasan ibn al-Haytham was born in Basra in around AD 965. Working in Cairo, the capital of the Fatimid Caliphate of north Africa, Ibn al-Haytham is now known mostly for his monumental *Kitāb al-Manāẓir* (Book of Optics). It is a remarkably modern-seeming work, with descriptions of experiments that include the apparatus, measurements, results, and conclusions, showing that Ibn al-Haytham worked in ways that a scientist today would recognize. He used the geometry and mathematics of Euclid to develop a detailed theory of sight involving the reflection, transmission, and refraction of light rays. Europeans called him by the Latinized name Alhazen, and in the thirteenth century much of what they knew about light and optics came from his book.

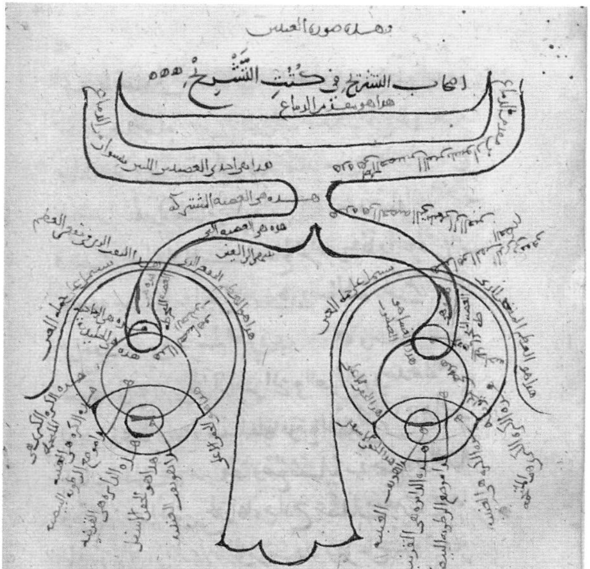

A diagram of the eyes and related nerves. From a copy of Ibn al-Haytham's *Kitāb al-Manāẓir*, eleventh century, MS Fatih 3212, Vol 1, Süleymaniye Manuscript Library, Istanbul, Turkey.

Ibn al-Haytham figured out the principles behind an optical device called the camera obscura that had been known since ancient times. It is basically a pinhole camera: light enters a box or chamber through a tiny hole in one side and falls onto a screen on the opposite side, where an image of the scene appears upside down. Ibn al-Haytham explained this flipping in terms of the paths light rays take as they pass through the pinhole: a clear statement of the fact that light travels through empty space in straight lines. Ibn al-Haytham also argued that as light is admitted into the hole—for example, by removing an obscuring curtain—it takes a certain time to reach the back wall. In other words, he implied, the speed of light is finite (but far too fast for him to measure).

Ibn al-Haytham's ideas were constantly guided by experiment. He made a camera obscura and used it—as people still do today—to view a partial solar eclipse, creating an image of the sickle shape of the Sun partly obscured by the Moon. It seems likely that his device was no mere box but rather an entire room or chamber (he used the term *al-Bayt al-Muẓlim,* meaning "dark room") in which images were projected onto the back wall.

The Book of Optics is filled with detailed descriptions of experimental procedures: "Let the

IBN AL-HAYTHAM
CA. 965–CA. 1040

Abū 'Alī al-Ḥasan ibn al-Haytham's education suggests that he was from a well-to-do family, and it may be that his reputation as a mathematical genius led the Fatimid caliph to offer him a court post in Cairo in around 1010. Legend has it that Ibn al-Haytham immediately over-reached himself by offering to build for the caliph a great dam on the Nile that would control its floods, only to then realize that it was far too ambitious a scheme. To escape the caliph's anger, Ibn al-Haytham is said to have feigned madness and been kept under house arrest until the caliph died in 1021, liberating the scholar to pursue his studies. Such tales add color to the history of science, but for precisely that reason must be taken with a pinch of salt.

See also: Experiment 32, Modeling the rainbow, early 14th century (page 140); Experiment 33, The origin of colors, 1666 (page 144).

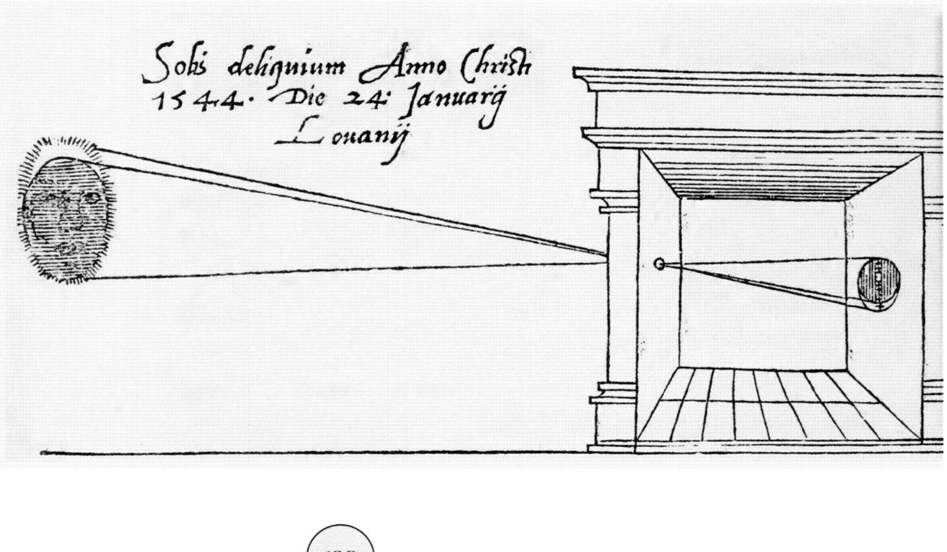

Illustration of an eclipse of the Sun, as observed on January 24, 1544 by Gemma Frisius and thought to be the earliest known representation of a camera obscura. From Frisius's *De Radio Astronomica et Geometrica Liber*, Antwerp: G. Bontius and Louvain: P. Phalesius, 1545, John Brown Carter Library, Rhode Island.

experimenter use a chamber into which sunlight enters through a wide hole of a magnitude not less than one cubit by the same, and let the light reach the chamber's floor…" From such careful observations, Ibn al-Haytham said, one might advance to general hypotheses: "we should ascend in our enquiry and reasoning, gradually and orderly, criticizing premises and exercising caution in regard to conclusions." It was a flawless description of how experimental science may lead to an understanding of natural phenomena.

Ibn al-Haytham's ideas informed theories of vision over the next several centuries. The eye came to be seen as a kind of optical device, rather like a camera obscura, with the retina as a screen and a lens to focus the light. Curiously, although he wrote widely on the mechanisms of vision, he did not make that connection—with the corollary that the image is cast onto the retina upside down and needs to be inverted by the brain. Still, his work set new standards for investigating nature based on experience and systematic observation.

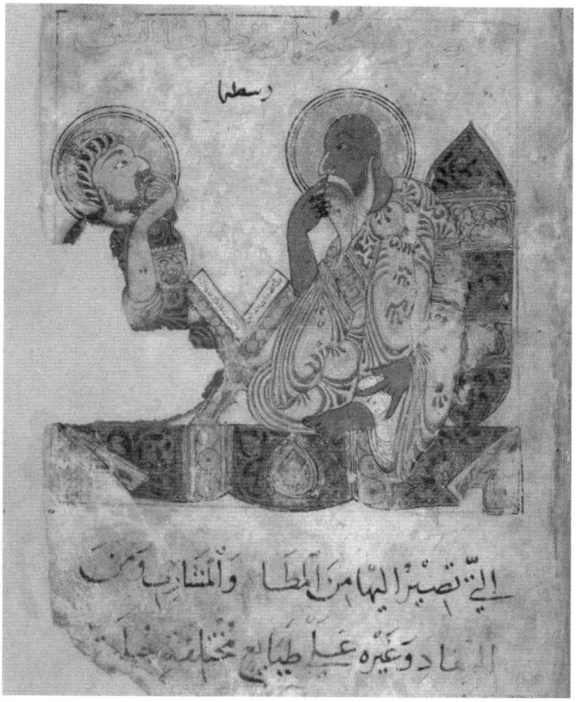

Above: Illustration of Aristotle teaching. From Jabril ibn Bukhtishu's *Kitāb na't al-Ḥayawān* (Book of the Characteristics of Animals), ca. 1220, The British Library, London.

Below: Leonardo da Vinci's ink on parchment drawing showing the similarities between the eye and a camera obscura. From da Vinci's *Codex Atlanticus*: Italy, 1478–1519, Ambrosian Library, Milan.

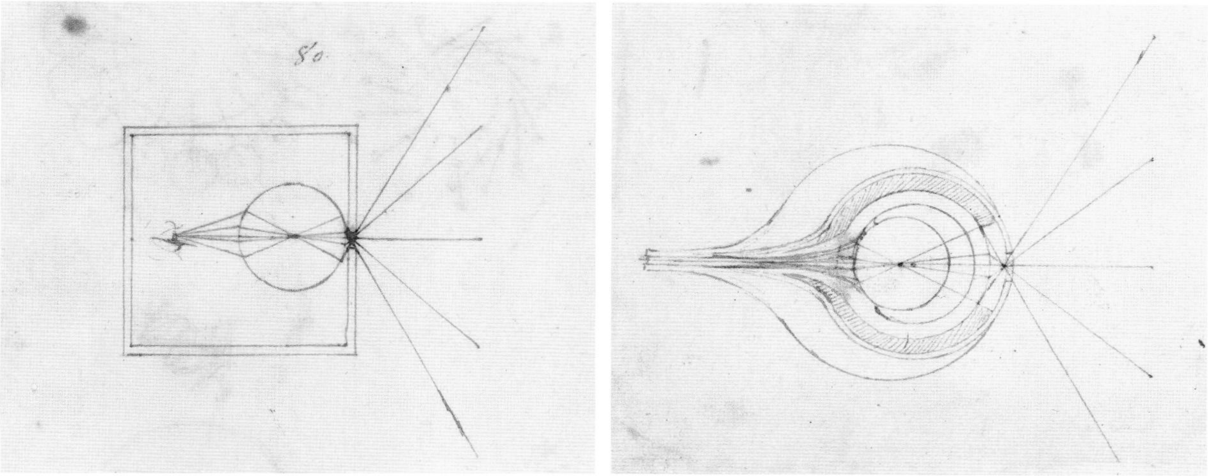

32

Modeling the rainbow (early 14th century)

Q How does a rainbow form?

Of all the natural manifestations of optical phenomena, the rainbow was long regarded as one of the most glorious. Its symbolism is universal: it is God's sign of "a covenant between me and the earth," as he tells Noah after the Flood; it is the bridge between the earthly realm and Valhalla in Norse mythology; and to the Greeks it was personified by Iris, the messenger of the Olympian gods.

Yet to a committed naturalist like Aristotle, it was also a phenomenon to be explained by reason and logic: by physics. In his book *Meteorologica* he argues that the rainbow is caused by sunlight being reflected from clouds as if by a mirror. That, of course, leaves several questions hanging. Why an arc? Whence the colors?

It was widely thought that light is altered when it interacts with matter, so that white sunlight passing through a prism becomes "stained" with the familiar spectrum running from red through yellow and green to blue and purple. (Exactly

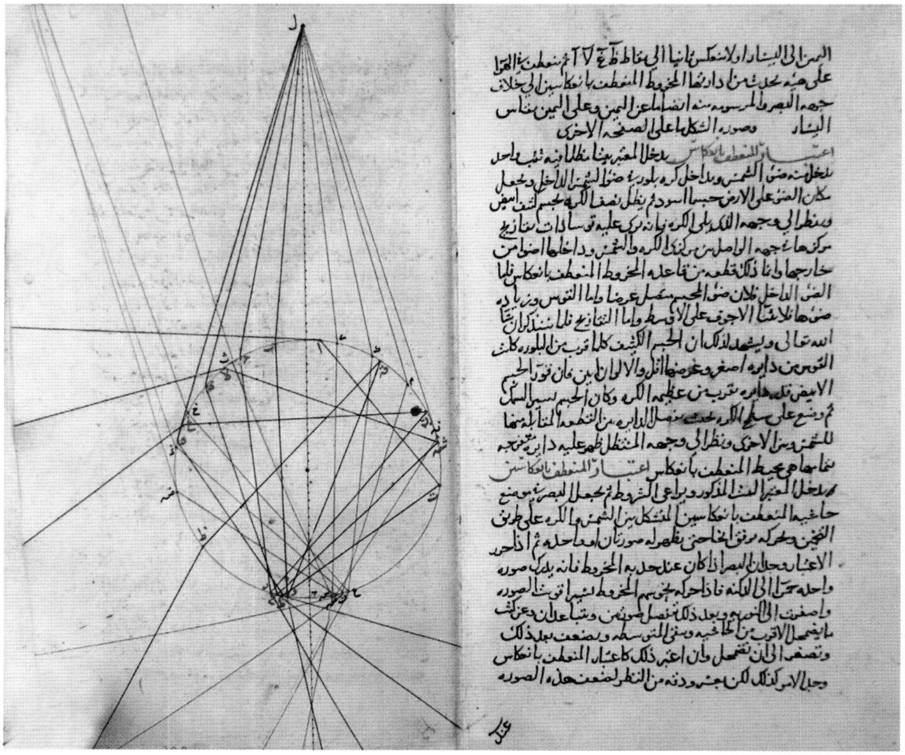

Diagram from Kamāl al-Dīn al-Fārisī's autograph manuscript of *Tanqih al-Manazir* (The Revision of Ibn al-Haytham's Optics), 1309, Adilnor Collection, Malmö, Sweden.

what are the principal colors of the rainbow was a matter of debate; in medieval art it is often depicted in bands of red, yellow, and green, echoing the Holy Trinity.) Some suggested that the rainbow's circular form was perhaps a reflection of the shape of the Sun itself.

The rainbow inevitably came within Ibn al-Haytham's purview (see page 137), and his partial theory invoking reflection was improved by the Persian mathematician Kamāl al-Dīn al-Fārisī in the early fourteenth century. In the West, the rainbow was studied by several eminent scholars in the thirteenth century, including Albertus Magnus in Germany and Roger Bacon and Robert Grosseteste in England. None of them cracked the problem, but the biggest advance during the Middle Ages came from a Dominican cleric named Theodoric of Freiberg in the early fourteenth century. In 1304, Theodoric tells us, he was asked by the Master of his order to write down what he had learned about this natural phenomenon, implying that he had already been studying it for some time. The result was Theodoric's *De iride et radialibus impressionibus* (On the Rainbow and the Impressions Created by Irradiance) written sometime between 1304 and 1310. It is one of the finest testaments to the intellectual vigor of the Middle Ages, challenging the common caricature that all medieval scholars did was to unthinkingly regurgitate the ideas of Aristotle adapted to the mold of Christian dogma.

Theodoric pays due respect to Aristotle's views on the rainbow, but then uses the Greek's own words to justify dissenting from them—for hadn't Aristotle said that "one never should depart from that which is evident from the senses"? It is an invitation to empiricism and experiment: to look for yourself rather than just accepting established authorities. True knowledge, Theodoric says, comes from "the combination of various infallible experiments with the efficacy of reasoning."

Unsurprisingly, some of that reasoning looks perplexing to us today. Theodoric claimed that

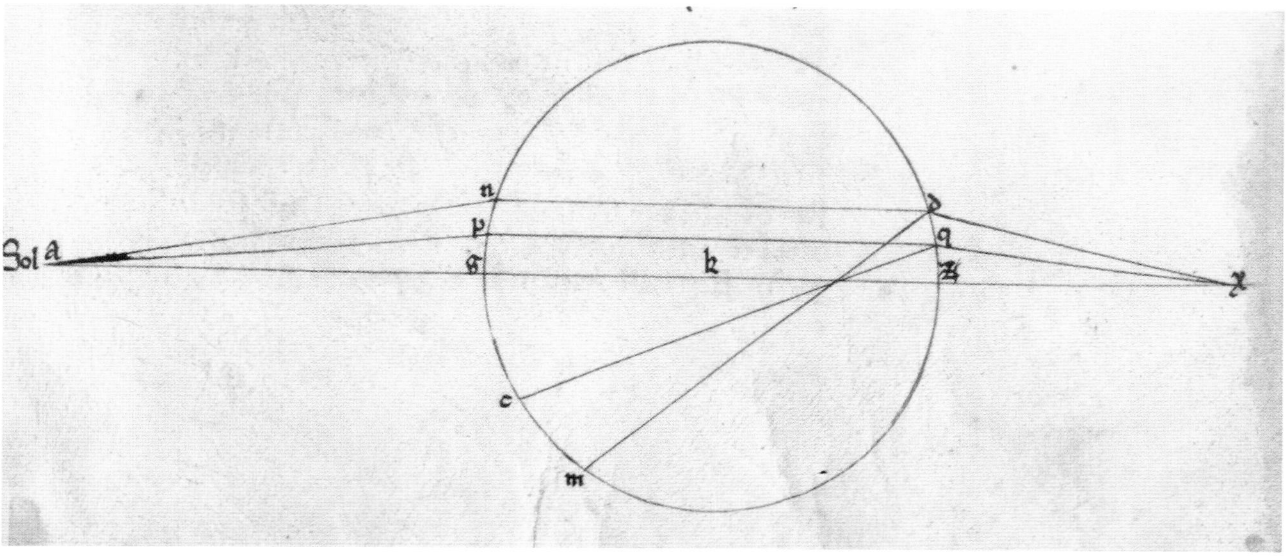

Theodoric's explanation of the rainbow for the first time invoked both the processes of refraction and reflection when light rays pass through a raindrop. From Theodoric of Freiberg's *De iride et radialibus impressionibus,* III, 2, 5, In: Manuscript Basel F. IV.30, University of Basel, Switzerland.

THEODORIC OF FREIBERG
CA. 1250–CA. 1310

In Theodoric of Freiberg we can find a perfect riposte to the suggestion that the medieval church opposed and suppressed curiosity about the natural world. It was around the time Theodoric studied in Paris that a notorious battle of authority was played out in the city between the laws of nature, as stated by Aristotle, and the omnipotence of God. But the dispute evidently did not deter the devout Dominican Theodoric from his studies of light, nor from boldly claiming to have found new knowledge. The mystery is more how he found the time between his heavy administrative duties for the church.

See also: Experiment 31, The camera obscura, early 11th century (page 137); Experiment 33, The origin of colors, 1666 (page 144).

there are four colors in the rainbow: red, yellow, green, and blue. But rather than seeing these as phenomena to be explained, he makes them almost axiomatic and treats them in the manner characteristic of ancient philosophy: as categories to be arranged into a system. Aristotle considered colors to be organized as contrary pairs, ultimately generated by the duality of light and dark. For Theodoric, red and yellow are "light" or clear colors that depend on how "bounded" is the medium in which they arise. Green and blue are "dark" or obscure colors produced by the medium's opacity.

It seems a rather ad-hoc system. But Theodoric figured he could then explain the order in which these colors appear in the rainbow according to what happens to the Sun's rays as they strike a raindrop. Specifically, he made the crucial step of considering not just reflection from the drops, as Aristotle had, nor solely refraction, as Albertus Magnus and Robert Grosseteste had, but both. In an illustration in *De iride* that is both remarkably modern and basically correct, Theodoric shows the rays entering the sun-facing front of a drop, being refracted as they pass through, then reflected back from the inner surface at the rear, passing again through the drop and being refracted as they exit. In this way, the rays travel from the Sun into and out of the raindrops to reach the eye of the observer. Depending on how deeply they travel away from the "boundedness" of the drop's surface, they take on different colors. Rather than the cloud as a whole acting as a mirror, each raindrop is its own little prism and reflector.

What Theodoric gets right is the way that different paths, all comprising two refractions and a reflection, correspond to different colors. What's more, he explains the circular shape of the bow based on the circular symmetry of the whole sun-drop-observer system: nothing changes if you rotate it. And he can explain the occasional appearance of a secondary bow, with an inverted sequence of colors, on the basis of double reflections from the inner surface of the drops.

But he is wrong about how the colors are produced. And he gets into geometric difficulties by assuming that the Sun and the observer are at similar distances from the raindrops, so that the Sun's rays are diverging rather than arriving essentially parallel to one another. He was also strangely wrong about the geometric characteristics of the rainbow, stating that the largest angle of elevation of its highest point above the horizon is 22 degrees, whereas it was already well recognized that it is almost twice that. The obvious inconsistency has to make us wonder how strongly felt in Theodoric's day was a need for agreement between theory and observation.

The crucial point, however, was that he didn't just conclude all this by deduction: he studied it experimentally. He *modeled* raindrops (as we'd now express it) using flasks with spherical bodies, filled with water—quite possibly those used by doctors to collect and inspect a patient's urine. This

Meteorology: A Double Rainbow. Color lithograph by René-Henri Digeon after Etienne Ronjat, 1868, Wellcome Collection, London.

is a remarkable gambit. It wasn't clear that water droplets in clouds are spheres, and they are, of course, much tinier than flasks—so it's not obvious that the latter are a good experimental proxy. But that's what Theodoric assumed, and it reflects experimental practice today: to represent some complex natural system with a simple, idealized analog that you can conveniently study in the lab.

Theodoric's explanation for the rainbow remained the best available for over three centuries. When René Descartes considered rainbows in his 1637 book *Les Météores*, his account is barely any different, even to the extent of experimenting with glass globes. But he makes no mention of Theodoric—a failure to give due credit that he would never be allowed to get away with today.

33

The origin of colors (1666)

Q Why does a prism transform light into a multicolored spectrum?

The puzzle of the rainbow was resolved in the seventeenth century through the work of the scientist who some regard as the greatest ever to have lived. In 1666, Isaac Newton—then a 23-year-old Cambridge graduate—performed an experiment with light that transformed our understanding of it. While it was thought that the bar of rainbow colors—called a spectrum—produced when white light (like sunlight) travels through a glass prism is caused by some property of the prism that alters the light, Newton showed the colors are already inherent in the light itself.

Legend has it that Newton did the experiment at his family home in Woolsthorpe, Lincolnshire, to which he had returned to escape the Great Plague that ravaged England in 1665. It did not, after all, require any fancy apparatus—just a few prisms, which could be bought almost as trinkets at markets (although he needed good-quality ones!). While there's truth in that, Newton had been planning such experiments for a while in his Cambridge room: we need not credit the plague for stimulating this leap in understanding optics.

Newton didn't report his results until six years later, when he sent an account to the Royal Society in London, the intellectual center of "experimental philosophy" in the mid-century. He was famously reluctant to disclose the outcomes of his studies, and had to be cajoled into writing down his celebrated laws of motion and theories about the motions of the planets in his masterwork the *Principia Mathematica* in 1687. The book in which he recorded his experiments and theories about

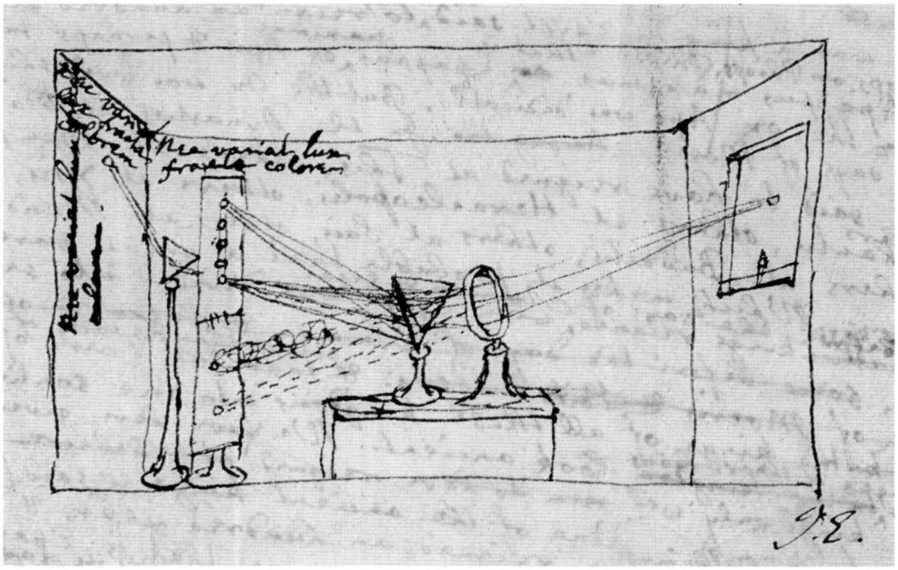

Isaac Newton's pen and ink sketch of his *experimentum crucis*, in which he refracted sunlight using a camera obscura and two prisms, ca. 1666, MS 361/2, New College Library, University of Oxford.

light, *Opticks*, was finally published in 1704. This was not so much because Newton was diffident about his work; on the contrary, he was rather covetous about it, and highly sensitive to criticism.

Newton begins his 1672 account by relating his surprise that the colored spectrum produced by his prism was rectangular in shape rather than circular, "as the received laws of Refraction" would lead one to expect. It seems a rather trifling question, especially to lead to such profound conclusions. In fact, his "surprise" is hard to credit, for this effect of a prism was well known, not least to Newton himself, who had been fascinated with such instruments since he was a boy. Newton was here no doubt indulging what is now a common practice in scientific papers: to construct a retrospective story so as to give a comprehensible narrative arc to a description of experiments that might have a more haphazard genesis and perhaps initially a different goal entirely.

At any rate, Newton embarked on a thorough program of experimentation to figure out what the prism was doing to light. One can imagine him almost literally playing with prisms, screens, and lenses until he found a configuration that allowed him to formulate and investigate some definite hypotheses. (Newton once famously claimed that "I feign no hypotheses," but in truth one can hardly do science at all without them.) It's a common situation for experimental science: you might want to investigate a phenomenon but be unsure quite what the right questions are, let alone how to deploy your instruments and measuring devices to answer them. You need to develop a *feeling* for the system you're trying to study.

Newton closed the "window-shuts" of his room, admitting a single narrow beam of sunlight through a hole, which passed into the prism. In the crucial experiment, Newton investigated the nature of the light *after* it exited the prism. If the light became colored because of some transformation produced by the prism, then a passage through a second prism might be expected to alter the light again. Newton used a board with a hole in it to screen off all the spectrum except for a single color—red, say—and then allowed that colored light to pass

ISAAC NEWTON | 1642–1727

John Maynard Keynes's famous remark that Isaac Newton was not the first true scientist but the "last of the magicians" captures the struggle historians have had to comprehend the man who some consider the most illustrious scientist of all time. Newton was born in Lincolnshire on Christmas Day 1642, and was a prodigy by any standards. He began his studies at Trinity College Cambridge in 1661; eight years later, aged twenty-seven, he held the prestigious chair of Lucasian Professor of mathematics. Newton's work gave rise to the notion, which still largely persists today, that all of the physical world can be explained on the basis of underlying mathematical laws governing the movement and interactions of particles. In the eighteenth century this picture led some scholars to call for a kind of secular "Religion of Newton." Yet Newton remained profoundly religious, if in an unorthodox manner. He denied the Holy Trinity and spent the later years of his life trying to establish a chronology of the Old Testament and to understand its apocalyptic prophecies. Moreover, he firmly believed in the alchemical transformation of metals, at a time when many were abandoning that old idea. Having revolutionized science—for once the word is apt—with his treatise on mechanics known as the *Principia Mathematica* (1687) and then his *Opticks* (1704), Newton mostly abandoned his scientific research after becoming a Member of Parliament for Cambridge in 1689 and Master of the Royal Mint in 1699. He reminds us how unwise it is to suppose that what it can mean to be a "scientist" in former times must conform to our expectations today.

See also: Experiment 32, Modeling the rainbow, early 14th century (page 140); Experiment 34, The wave nature of light, 1802 (page 150).

through the second prism. He found that this light emerged from the second prism refracted—bent at an angle—but otherwise unchanged.

In other words, a prism seems only to bend (refract) light, leaving it otherwise unaltered. But it does so to different degrees (that is, at different angles) for different colors. This in itself was nothing new: the Anglo-Irish scientist Robert Boyle had said as much in his 1664 book *Experiments and Considerations Touching Colours*, which Newton had read. But only Newton saw what this implies: that refraction is then all there is to it. The colors themselves are already in the white light, and all the prism does is to separate them out. As he put it, "*Light* consists of Rays differently refrangible" [meaning *refractable*]. The colors of the spectrum, then, "are not *Qualifications* [alterations] *of Light* …

Spectrum formed by white light through a prism. From Martin Frobenius Ledermüller's *Drittes Funfzig seiner Mikroskopischen Gemüths- und Augen-Ergötzungen*, 1762, Vol. II, Plate II, Wellcome Collection, London.

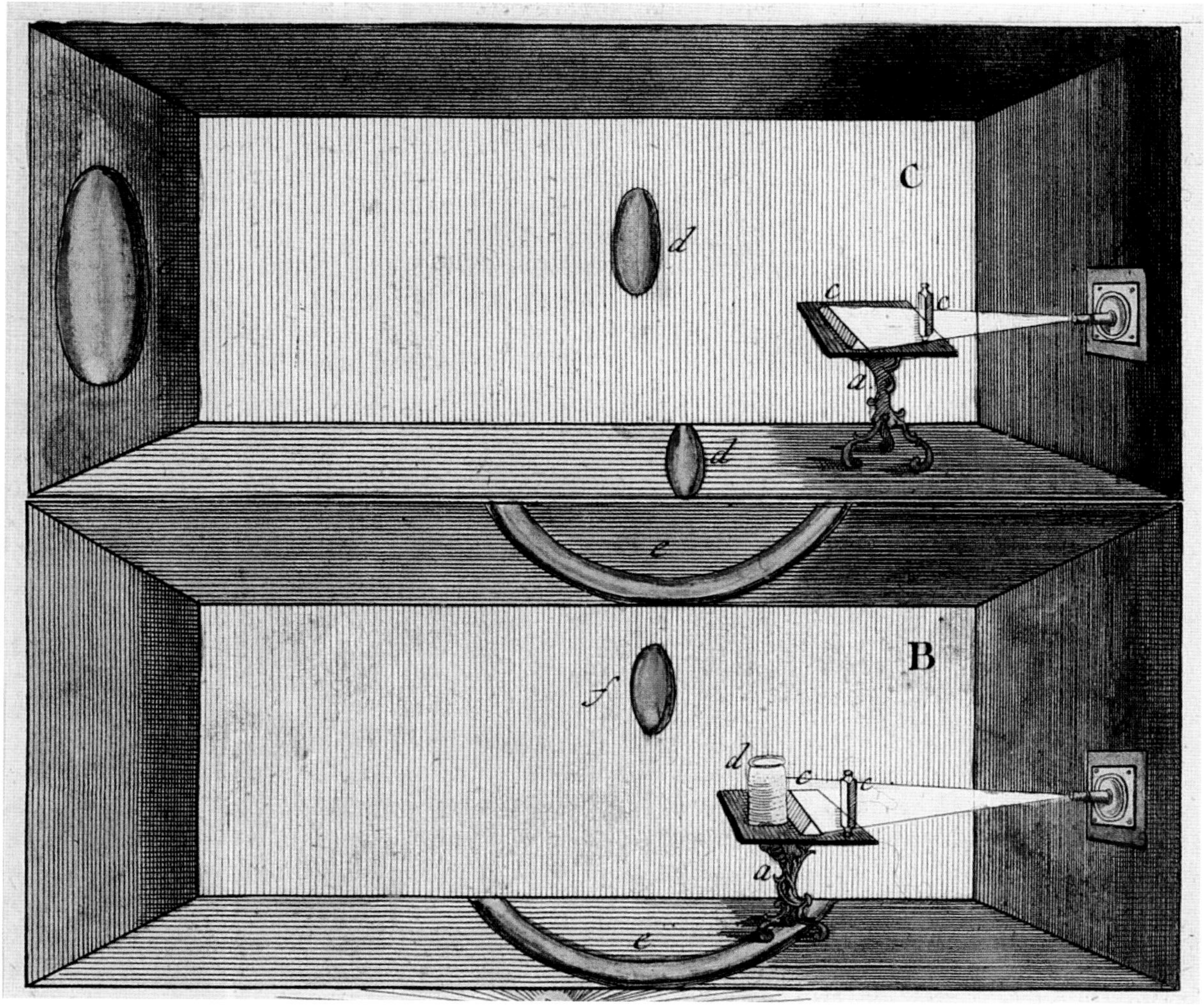

Photograph showing white light dispersed through a prism and split into the colors of the rainbow.

(as 'tis generally believed), but *Original* and *connate properties*."

That was a bold interpretation: sunlight was not, so to speak, elemental, but compound. To test this idea, Newton used a lens to refocus a many-hued spectrum into a single, merged beam—which, he observed, was white. He also passed this reconstituted beam through another prism to reveal that it could again be split into a spectrum just as before.

Newton explained how his observations could account for the rainbow, produced by the refraction and reflection of light through raindrops that act as tiny prisms. The colors of everyday objects, he added, arise because they reflect "one sort of light in greater plenty than another." And the results explained the defects of lenses (Newton himself had become adept at making these by grinding glass), whereby refraction of different colors produces a defocusing effect called chromatic aberration.

The Royal Society's secretary Henry Oldenburg told Newton that his report was met with "uncommon applause" when read at a gathering in February 1672. But not everyone appreciated it. After the paper was published in the society's *Philosophical Transactions*, its in-house curator of experiments, Robert Hooke, who considered himself an expert on optics, presented several criticisms (which we can now see were mistaken). Newton replied with lofty condescension, igniting a long-standing feud between the two men. One problem is that Newton's experiments, despite their apparent simplicity, are not easy to replicate: some, in England and abroad, tried and failed. But they have stood the test of time, a testament to the power of experiment to literally illuminate the unknown that, in the judgment of philosopher of science Robert Crease, gives Newton's so-called *experimentum crucis* "a kind of moral beauty."

Sir Godfrey Kneller's oil on canvas portrait of Sir Isaac Newton, 1689, from a private collection.

The art of scientific instrumentation

In 1664 the English diarist Samuel Pepys visited the London store of Richard Reeve, who was considered the finest maker of scientific instruments in the country. There he paid the "great price" of five pounds and ten shillings for a microscope—the new must-have device for any gentleman who, like Pepys, maintained an interest in natural philosophy (see page 34). Pepys pronounced it "a most curious bauble," but at first he couldn't see much with it—even with today's instruments there is a certain amount of "getting your eye in" to be able to use a light microscope well.

It might seem odd that a dilettante such as Pepys (who even became president of the Royal Society in 1686, although he did no research of his own) would dash to the store and buy the latest scientific instrument. You would hardly expect an affluent intellectual today to hanker after a nuclear magnetic resonance spectrometer. But in the seventeenth century, before science was professionalized, craftspeople like Reeve could rely on a clientele from among the wealthy classes. This was reflected in their wares: illustrations of microscopes of that time show devices embellished with fancy engraving that has no practical purpose whatsoever. These instruments were built to look good on the shelf, and were objects of desire and prestige. As historian of science Catherine Wilson puts it, one might suspect that "the scientific community in the seventeenth and eighteenth centuries was more interested in designing and possessing optical instruments than in using them to explore the world."

This association of scientific instruments with social status in the formative stages of modern science might partly help to explain why

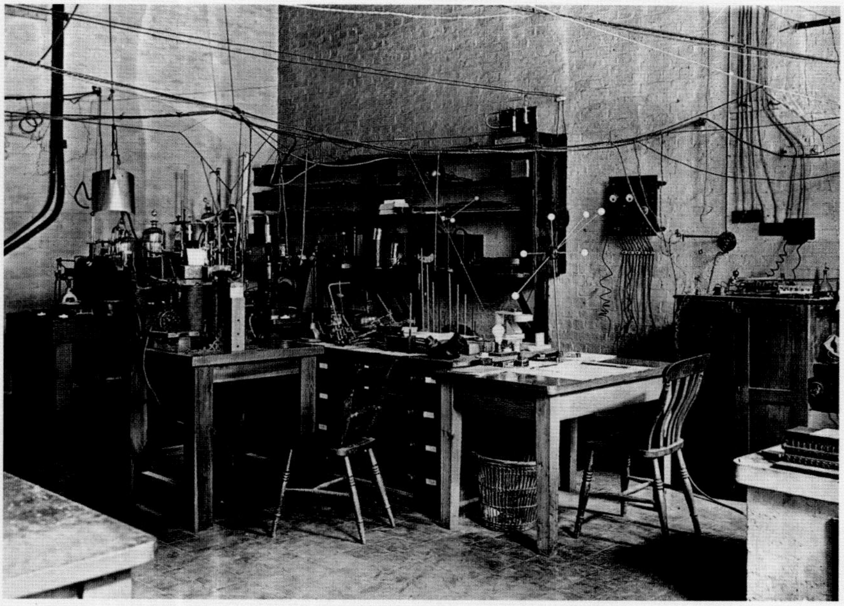

Ernest Rutherford's research laboratory at the University of Cambridge, 1915.

the apparatus of that era seems so exquisitely manufactured and peculiarly beautiful, made not from the utilitarian steel and plastic of modern machines but from luxury materials like brass and mahogany. Historian of technology Derek de Solla Price has argued that instruments in former times typically went through a decorative and symbolic phase before settling into a more utilitarian form. Some people collect these antique scientific instruments now as if they were works of art; some of them surely are.

But the care lavished on the aesthetic qualities of those instruments was not just a question of sales; it was also about establishing authority. The air pump made by Robert Hooke and instrument-maker Ralph Greatorex for Robert Boyle's experiments on vacuums in the 1650s (see page 48) was beyond the means of most other scientists in Europe, and the image of it in Boyle's treatise is as beautiful as the device itself surely was, all swelling tubes and sprouting stopcocks lovingly drawn in careful perspective and dramatic chiaroscuro. The message it conveys is that the results one can obtain with such a splendid device are far more trustworthy than anything gleaned from an instrument cobbled together from whatever was to hand. Boyle's air pump features in the elaborate frontispiece of Thomas Sprat's 1667 *History of the Royal Society*, symbolizing both the ingenuity and the authority of the experimental philosophy. Like the awe-inspiring particle colliders of today, such unique and intricate devices could reveal secrets of nature that remain far beyond the reach of more prosaic technologies.

The same might be said of the scientific glassware of that age, which matches in elegance anything to be seen in artifacts intended for purely ornamental purposes. Glassblowing—a central skill for many different areas of science—has been particularly revered by experimentalists, some of whom cultivated the art themselves. And any technical glassblower worth their salt has at least a streak of the artist in them.

From aesthetics to functional practicality?

The contrast with some of the experimental apparatus of the modern era is stark. Photographs of Cambridge's Cavendish Laboratory when Ernest Rutherford worked there at the end of the nineteenth century make it hard to believe that this is where great discoveries in atomic science were made. The equipment looks like junk gathered on an old kitchen table in a garage. Rutherford himself was notorious for eschewing expensive equipment in favor of homemade apparatus seemingly made of old bits of tin, its extemporized nature perhaps reflecting the pattern of his theories. There is an improvised, clunky feel, too, to the first transistor cobbled together from a chunk of germanium at AT&T Bell Laboratories in 1947, and the first scanning tunneling microscope made in 1981 (see page 116) looked far too makeshift an affair to attain the finesse needed for "seeing" atoms. Now that science was professionalized, there was no audience to impress in day-to-day experimentation (though the equipment might be smartened up for the press).

Despite their funding awards, few scientists have the wealth of a Boyle to lavish on their equipment in any case. Today's methods of making scientific instruments exploit new possibilities, such as 3D printing and robotic automation, to make apparatus that is cheap and compact. Yet this doesn't mean that aesthetics are irrelevant for modern instruments. In some ways, the aesthetic has simply changed to reflect the image modern science wishes to project: impersonal and sober (no bright colors!) but also sleek and efficient. Anyone who spends time with experimentalists comes to realize that they still see instruments as sources of prestige and gratification: there is elation when the new state-of-the-art microscope or spectrometer arrives, whereas the old one—well, we're fond of it, but honestly it looks ***so*** unfashionable now. The first thing a scientist wants to do with a visitor is to show them the lab: instruments are a source of pride as well as knowledge.

34

The wave nature of light (1802)

Q Is light made of particles or waves?

Among the sources of bitter dispute between Isaac Newton and Robert Hooke in the seventeenth century was the nature of light. Newton was convinced that it consisted of particles—"corpuscles"—whose differences in mass or speed accounted for the different colors. Hooke maintained that light is "nothing but a pulse or a motion propagated through an homogeneous, uniform, and transparent medium"—in other words, a wave.

Most natural philosophers sided with Newton. That was generally wise in scientific matters, but not on this occasion. Ironically, Newton himself unwittingly produced evidence of the waviness of light. He reported that when a lens was placed, curved face down, on a flat glass plate, bright and dark rings (later called fringes) can be seen encircling the point of contact. These are caused by light waves being reflected off the glass surfaces in the narrow gap between lens and plate, and interfering with one another. Where the peaks of the waves coincide, they amplify each other and enhance the brightness, which is called constructive interference. Where a peak meets a trough, there is destructive interference, creating dark regions. Only a wavelike nature can account for this; to compound the irony, Newton also explained the phenomenon of interference in experiments on water waves.

But it was not until the start of the nineteenth century that compelling evidence was adduced for light's waviness. In 1800 the English polymath Thomas Young explained how interference can arise in sound waves, and in the following year he suggested that similar effects in light might explain Newton's optical rings. He drew an analogy with water waves, asking the reader to imagine two sets of waves exiting a lake into a narrow channel. "If they enter the channel in such a manner that the elevations of one series coincide with those of the other," he wrote, "they must together produce a series of greater joint elevations." But if they are perfectly out of step,

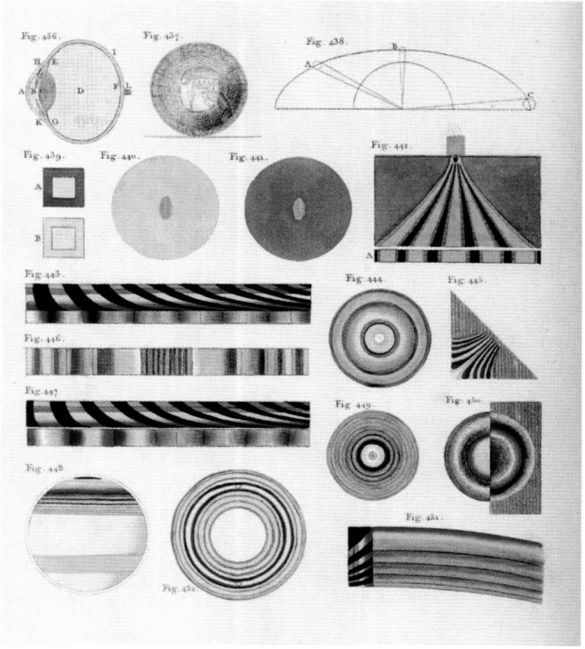

A range of optical effects, including the interference pattern. From Thomas Young's *A Course of Lectures on Natural Philosophy and the Mechanical Arts*, London: Printed for J. Johnson, 1807, Vol. 1, Plate XXX, University of California Libraries.

"the surface of the water must remain smooth." If light consists of waves, the same should happen when light beams cross.

In 1803 Young presented to the Royal Society a paper describing experiments that revealed such interference. He sealed off a room from light as Newton had done, and then admitted a sunbeam through a needle-hole in a piece of paper. Placing a very narrow strip of card into the beam, he saw colored stripes ("fringes") on either side of the shadow, which he interpreted as interference patterns from the light passing on either side. The colors arise because different wavelengths of light interfere at different positions.

All the same, Young was circumspect with his conclusions, not mentioning waves but saying only that the nature of light must bear a "strong resemblance" to that of sound. But his target was clear: he added that "those who are attached to the Newtonian [that is, corpuscular] theory of light ... would do well to endeavour to imagine any thing like an explanation of these experiments, derived from their own doctrines."

And Young had stronger experimental evidence that light consisted of waves, which he presented during a series of lectures at the Royal Institution in 1802 and 1803. First he showed how two sets of waves radiating in a tank of water will interfere to create a series of bands in just the manner he had adduced previously. Then he performed the same experiment with light, creating two sources by casting a beam onto "a screen in which there are two very small holes or slits," which act as sources that radiate in all directions on the far side. When that light falls onto another screen, it "is divided by dark stripes into portions nearly equal"—interference fringes.

If water waves are carried by water itself, and sound waves by air, what medium fluctuates to produce light waves? Young assumed that there must be some tenuous, invisible fluid called the ether in which the waves propagate.

Young published these findings only in 1807, and it seems that those aware of the lectures were not fully persuaded by them—Young was apparently a rather dry speaker, with none of the flair of the Royal Institution's star lecturer Humphry Davy. Davy himself seemed sceptical in a letter he sent to a correspondent in 1802, saying: "Have you seen the theory of my colleague, Dr. Young, on the undulations of an Ethereal medium as the cause of Light? It is not likely to be a popular hypothesis after what has been said by Newton concerning it." Some other attacks on Young's wave theory of light were more explicit and extreme. When he presented them to the Royal Society, one critic chauvinistically jeered "Has the Royal Society degraded its publications into bulletins of new and fashionable theories for the ladies of the Royal Institution?"

Young was not a good advocate for his ideas, and they caught on only slowly as others repeated his double-slit experiment. It's not as easy to do as it sounds, requiring rather precise and sharp-edged holes or slits, but with care, the interference fringes can be produced in the classroom today with a laser pointer as the light source. All the same, the "particle theory" of light was not killed off by the double-slit experiment, but was later resurrected in a new form.

THOMAS YOUNG | 1773–1829

Raised in a Quaker family, Thomas Young was a prodigy who could read at the age of two and was learning Latin by six. While training as a physician, he became interested in how the eye works, and he developed the first theory of our three-color vision system. In 1801 he gave up medicine to become a professor and lecturer at the newly formed Royal Institution in London. Sometimes called the last man who knew everything, Young studied how liquids spread, the mechanical properties of materials, and the principles of musical harmony, and was a polylinguist and a keen Egyptologist who helped to decode the Rosetta Stone and thereby decipher ancient hieroglyphics.

Measuring the speed of light (1849)

Q How fast does light travel through space?

Because light travels so fast, it was generally assumed before the seventeenth century to move with infinite speed—that is, to be a phenomenon experienced instantaneously. Galileo was one of the first to question that assumption by attempting to measure the speed of light. Stationing himself on a hilltop with a lamp, and an assistant on a distant hilltop with another lamp, he uncovered his own lamp and instructed the assistant to uncover his as soon as Galileo's became visible. The method was far too crude to deliver a realistic measure, but it established the idea that, while "extraordinarily rapid," light did not arrive instantaneously from a distant source.

A much better estimate was made in 1676 by the Danish astronomer Ole Rømer, based on the timing differences of eclipses of the four major moons of Jupiter, depending on the relative positions of the Earth and Jupiter along their orbits: the eclipses seem to happen later when the two planets are farther apart. There is a more significant delay in travel time over such astronomical distances, and so Rømer was able to deduce a rather good estimate of 240,000 kilometers per second: around 80 percent of the

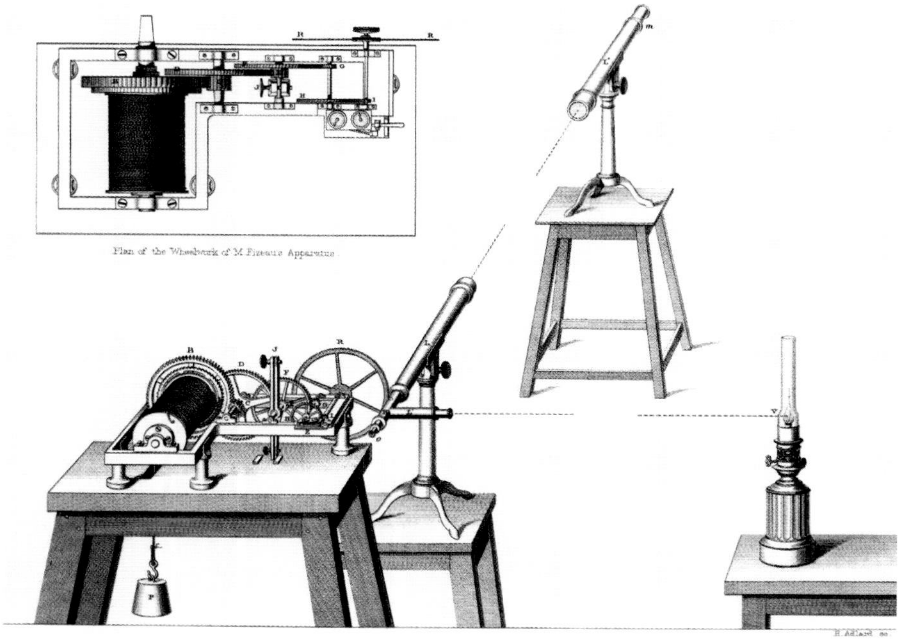

Armand-Hippolyte-Louis Fizeau's apparatus for measuring the speed of light by sending a beam back and forth from a mirror between the teeth of a rapidly spinning cogwheel. From *Astronomie Populaire*, 1858, Vol. IV, Fig. 339, Royal Astronomical Society, London.

accepted value today of 300,000 km/s. In 1728, English physicist James Bradley used another astronomical method to make a better estimate of 301,000 km/s.

In 1849 the French scientist Armand-Hippolyte-Louis Fizeau devised an ingenious new approach to the problem. He was working at the Paris Observatory in collaboration with Léon Foucault, a friend from his days as a medical student. The observatory director suggested that the pair try to measure the speed of light using an Earth-based experiment. After working together on the matter for a short time, they decided each to devise their own method.

Fizeau's scheme involved sending a beam of light through a gap in a fast-spinning cogwheel and then, reflected from a mirror, back again. If, in the brief instant it took the light to travel from the cogwheel to the mirror and back, the wheel had rotated so that a cog was now positioned where the gap originally was, the beam would be blocked. Knowing the rotation speed of the wheel and the distance to the mirror, one could work out the light beam's speed.

That summer, Fizeau set up the apparatus at his family house in Suresnes, near Paris (now a suburb of the city). A lamp was focused by a lens into a beam that passed through the cogwheel and traveled 8.6 kilometers (5½ miles) to a mirror on the hill in Montmartre and back, where it could be seen by the observer if it passed again through the wheel's teeth. From this experiment Fizeau calculated a speed of 313,300 km/s—within 5 percent of the actual value, which was not bad for such a simple mechanical experiment. Foucault made improvements to the experimental design, replacing the cogwheel with a rotating mirror, and in this way he improved the estimate in 1862 to 299,792 km/s.

Light speed is immense on the human scale but may seem positively sluggish on cosmic scales. It takes eight minutes for light from the Sun to reach us, and more than four years for it to arrive from the nearest star. The light we see today from some distant galaxies set out well before our solar system was formed. This is why light years—the distance traveled by light in an earth year—are a convenient unit for measuring cosmic distances—and why, when we look out into the universe over huge distances, we are also looking back in time.

Fizeau went on to show that the speed of light in a flowing liquid did not change due to the liquid's movement—unlike the way a ball thrown on a moving train has the speed of the train added to that of the throw. This puzzling finding was not explained until Albert Einstein developed his theory of special relativity in 1905. Einstein's theory explained why the movement of the Earth through space, relative to the ether thought to carry light waves, makes no difference to light's speed. Fizeau himself had tried to detect such a difference in 1851, an attempt to which Einstein often referred.

ARMAND-HIPPOLYTE-LOUIS FIZEAU | 1819–1896

Born into a wealthy family, Fizeau was at leisure to pursue his scientific interests. After studying and then abandoning medicine, he collaborated with Léon Foucault to adapt photography for recording astronomical observations. This led François Arago, head of the Paris Observatory, to invite the pair to make the first ever photographs of the Sun, and then to measure the speed of light. Fizeau's successes brought him honors and eminence, and in 1878 he became president of the physics section of the French Academy of Sciences.

See also: Experiment 2, Direct demonstration of the rotation of the Earth, 1851 (page 16); Experiment 3, Attempting to detect the ether, 1887 (page 18).

Light as particles, particles as waves: quantum theory

It is in the nature of science that satisfying explanations are destined to raise more questions than they solve. So it was with atomic theory, evolution by natural selection, and genetics—and with the wave theory of light too. It worked so well until at the start of the twentieth century the tidy distinction between waves and particles vanished. It became unclear what a "particle" meant: not, certainly, some tiny lump of matter, for in the subatomic world particles look like local intensifications of nebulous, spread-out entities called fields. By the same token, light—now considered to be a wavy disturbance of electromagnetic fields—may take on a particle-like character. Only experiments can reveal such awkward truths and compel us to continue revising our conception of the world.

The photoelectric effect (1899-1902)

Q Why do metals emit electrons when irradiated with light?

There is often no telling where an experimental trail will lead. That was surely the case when in the 1880s the German physicist Heinrich Hertz set out to test James Clerk Maxwell's theory, adduced in 1865, that light is an electromagnetic wave. Hertz wanted to see if he could induce such a wave from an electrical phenomenon like the discharge of a spark. He inadvertently found that the strength—experimentally, the length—of one spark could be influenced by the light emitted by another spark, specifically by the light in the ultraviolet part of the spectrum, just beyond the minimum wavelength registered by the eye.

Why was ultraviolet light affecting the discharge? In 1899 J. J. Thomson in Cambridge found that such radiation could cause the ejection of electrons—the negatively charged particles he had shown two years previously to be the constituents of cathode rays (see page 112) from metals. It seemed that the oscillating electromagnetic field of ultraviolet light disturbs electrons in the metal atoms, transferring its energy to these particles until eventually they are shaken loose. Electron emission stimulated by light became known as the photoelectric effect. Thomson showed that it caused a negatively charged electrode to gradually lose its charge as electrons escaped, thereby lowering its propensity to generate a spark discharge.

In 1902 the German scientist Philipp Lenard, working at the University of Heidelberg, observed a puzzling fact about the photoelectric effect. Lenard devised an experiment that allowed him to measure not just the number of electrons emitted but also their energies. Typically, the emitting plate was placed in a glass vacuum tube and irradiated with light, and the electrons emitted were drawn to a positively charged plate connected to the emitter in a circuit. The current flowing through the circuit was then a measure of the number of electrons emitted. But Lenard gave the collecting plate a slight negative charge, so that it repelled electrons. The particles could only get to the plate if they had enough kinetic energy—energy of motion—to overcome the repulsive force. So by changing the voltage applied to the collector plate, Lenard could deduce this energy.

Lenard found that the energy of the electrons increased as the wavelength of the light

PHILIPP LENARD | 1862–1947

Although an exceptional experimentalist, Philipp Lenard was not able to stay abreast of the theoretical developments in physics, especially quantum mechanics, that his studies helped to precipitate in the early twentieth century. By instinct a passionate nationalist, Lenard became an ardent supporter of Adolf Hitler's Nazi party in the 1920s, and he hoped that Hitler's rise to power in 1933 would enable him to seize control of German physics. He promoted the idea that "true" physics (*Deutsche physik*) was practiced by "Aryans" and was based solidly in experiment, in contrast to the abstract theoretical physics propagated by Jewish scientists like Einstein. After the Second World War, Lenard, then in poor health, was deemed too frail to stand trial at Nüremberg for being a Nazi collaborator.

decreased, because shorter-wavelength light has more energy. What's more, the dimmer the light, the fewer electrons were emitted. All that made intuitive sense. But Lenard expected that even for long-wavelength (say, red) light, electrons could gather enough energy to be emitted if the light was bright enough. In contrast, he found there was a rather sharp cut-off wavelength above which no electrons would be emitted, regardless of the light intensity. In other words, the intensity affected only the number but not the energies of the electrons—and above the threshold wavelength, no amount of light could induce photoelectric emission.

Lenard was a masterful experimentalist, and his findings seemed irrefutable. But they

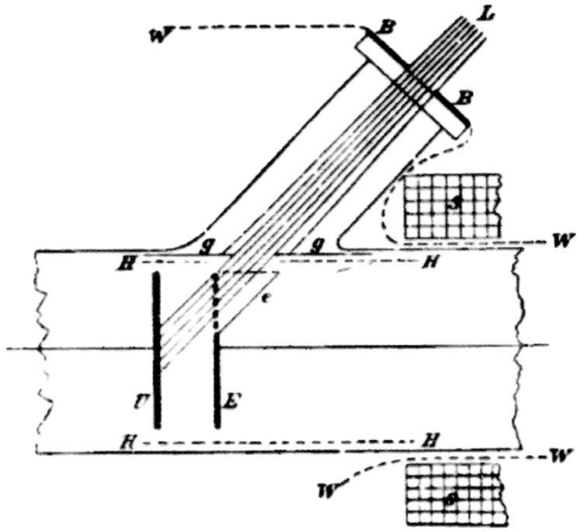

Below: English physicist Joseph John Thomson with the cathode ray tube that he used in his discovery of the electron in 1897, Cavendish Laboratory, Cambridge.

Above: Lenard's diagram of the photoelectric effect. From Philipp Lenard's "Über die lichtelektrische Wirkung," *Annalen der Physik*, 1902, Vol. 8, Issue 1.

made no sense in terms of classical physics, according to which the electrons in the metal plate could simply accumulate energy bit by bit from the light until they had enough to escape. In 1905 Albert Einstein proposed an answer. He drew on the notion, proposed five years previously by German physicist Max Planck, that particles vibrating in a material could only increase their energy of vibration in discrete steps, with the step size being proportional to the vibration frequency. Planck's law seemed to govern how electromagnetic waves are radiated from vibrating charged particles in a warm object. The energy steps were called *quanta*, and the vibrations were said to be quantized. This was in stark contrast to classical vibrations—of a spring, say—which can become more and more energetic in gradual increments of any size.

Planck had made his suggestion for the quantization of energies of atoms and subatomic particles as an ad-hoc way to explain experimental observations, without giving much thought to whether it was physically realistic. But Einstein proposed to take Planck's quantization literally: energies at these tiny scales really are quantized. What's more, he suggested that they apply to light itself. That's to say, light is composed of discrete "packets" of energy called quanta, each of which carries an energy proportional to its vibration frequency.

In the photoelectric effect, Einstein said, each electron that escapes has absorbed the energy of a photon from the incoming light. If we assume that there is a certain minimum energy needed to escape from the metal, then photons of light carrying energy quanta smaller than this threshold can never boost an electron enough to escape. As later experiments verified Einstein's theory, physicists became aware that they needed a new theory, different from classical mechanics, to account for how quantized particles behave. That emerged in the 1920s and was called quantum mechanics.

ALBERT EINSTEIN | 1879–1955

Albert Einstein, in the classic formulation, "needs no introduction." But it is curious that, for all the transformative impact his ideas had on modern physics, he was awarded the Nobel prize for the vague "services to theoretical physics"—and "especially for his discovery of the law of the photoelectric effect." Important though that was, it hardly hints at Einstein's foundational role in quantum mechanics or the impact of his theories of special and general relativity. This is often interpreted as caution on the part of the Nobel committee, who were even in 1921 not yet ready to fully accept such radical ideas. Einstein's 1905 paper on "light quanta" and the photoelectric effect was one of five published in that year of astonishing creativity, each of which transformed an aspect of physical science: a productivity that he was never able to repeat.

See also: Experiment 24, The nature of alpha particles and discovery of the atomic nucleus, 1908–1909 (page 108); Experiment 25, Measuring the charge on an electron, 1909–1913 (page 112).

Lenard considered himself the doyen of the photoelectric effect and was resentful that Einstein had seemingly explained something about the phenomenon that had foxed him. He rejected Einstein's ideas—not just in quantum physics but also relativity—and his antipathy curdled into virulent antisemitism as Germany descended into Nazi rule. His resentment only grew when Albert Einstein was awarded the 1921 Nobel prize in physics for his work on the photoelectric effect.

The diffraction of X-rays by crystals (1912)

Q How are X-rays scattered from crystals? What does the scattering pattern reveal?

The arguments about whether light is a wave or a stream of particles were recapitulated for the X-rays discovered by German physicist Wilhelm Röntgen in 1895. Some researchers suspected they were waves with a very short wavelength, of about the same scale as the size of atoms (the very existence of which was then still an open question). Others, including the English scientist William Bragg working in Australia, believed they might be particles rather like the alpha and beta particles of radioactivity. At any rate, X-rays and radioactivity were considered to be closely allied phenomena—the latter was, after all, discovered in 1896 in experiments on X-rays (see page 88). In 1906 the German physicist Arnold Sommerfeld at the University of Münich lamented in a letter, "Isn't it a shame that ten years after Röntgen's discovery one still does not know what is going on with X-rays."

However, Sommerfeld was determined to find out. In 1911 he invited Röntgen's former student Walter Friedrich to join his institute and conduct experiments on X-rays—rather to the chagrin of Röntgen, who had been counting on Friedrich's experimental assistance himself. Sommerfeld set Friedrich the task of investigating the X-rays emitted when cathode rays—beams of electrons—hit a target.

Friedrich, however, became diverted from Sommerfeld's task by another member of the Münich group, Max von Laue, who had studied with Max Planck in Berlin, where he had befriended Einstein. In early 1912, Laue went for a stroll with Paul Ewald, one of Sommerfeld's doctoral students, who was working on the way light interacts with crystals. Ewald happened to mention to Laue that the typical distance between atoms in crystals is about 1/500th to 1/1000th of the typical wavelength of light, which gave Laue an

Experimental apparatus used by Walter Friedrich and Paul Knipping in the discovery of X-ray diffraction, ca. 1912, Deutsches Museum, Munich.

idea—a "flash of inspiration," as he later put it—for how to test the possible wave nature of X-rays.

The interference effects that Thomas Young had described for water waves and light (see page 150) depend on having two or more sources of waves that are spaced apart at a distance roughly comparable to the wavelengths of the waves. Laue figured that such effects, giving rise to a pattern of light and dark regions, might occur for X-rays if they were to interact with atoms in a crystal, given that their wavelength was suspected to be comparable to the atoms' separation. This phenomenon of interference caused by the scattering of waves from or around an object is called diffraction.

While indeed we think of such X-ray diffraction today in terms of the scattering or reflection of the rays from atoms, that's not exactly how Laue saw it initially. He believed the atoms in a crystal would act not so much as scatterers but as sources of X-rays: he figured that they might absorb and then re-emit these rays. What's more, our modern view of crystals as orderly lattices of atoms was not then widely accepted either. Both Ewald and Laue later claimed that, outside Münich (where there were some influential advocates for the idea), rather few physicists shared this "lattice hypothesis" for the structure of crystals. That claim was challenged in 1969 by historian of science Paul Forman, who called it a myth and said the idea that crystals are regular stacks of atoms and molecules—dating back to the late eighteenth century—was already widely accepted.

So the true picture is rather confusing. Here was Laue, apparently suggesting that we find out what X-rays are by firing them at crystals, even though no one really knew what crystals are either. He didn't really know how the effect might arise, nor how to set up such an experiment, nor what exactly to look for. So he roped in Friedrich for the task. Despite Sommerfeld's scepticism that the experiment would work (oddly, given his eagerness to find support for the wave hypothesis), Friedrich agreed to give it a go—perhaps (he and Laue agreed) conducted at the back of the lab in the evenings. He enlisted in turn the help of another of Röntgen's students named Paul Knipping. To make it still more perplexing, such experiments in firing X-rays at crystals had already been tried by others, including Röntgen himself. But apparently they hadn't been thinking about diffraction effects, and so the fact that they saw none may have been because they weren't looking.

Even the nature of the experiments themselves is disputed. Where should the two researchers look for the scattered X-rays? Laue says that they placed photographic plates behind the target crystals (copper sulphate) and looked for transmitted rays. Friedrich and Ewald say that initially they looked for X-rays reflected in front of the crystal; Knipping seemingly suggested placing plates all around in order to cover the options. They barely knew what to expect: "we had no idea what it would look like," Friedrich later admitted.

At any rate, in April 1912 the two researchers did find that the plates were imprinted with dark spots in regular patterns, where bright beams of

MAX VON LAUE | 1879–1960

Max von Laue was typical of the physicists of his time in the breadth of his interests: from optics and solid-state physics to quantum theory and relativity. He was intimate with all the great physicists in Germany in the early part of the twentieth century, studying under Max Planck and Arnold Sommerfeld as well as befriending Einstein. He was also one of the most courageous among his peers in openly opposing the Nazi regime in the 1930s, decrying the persecution of Jewish scientists and opposing the ambitions of the Nazi sympathizer Johannes Stark to dominate German physics. It was said that he would carry parcels under both arms when he left his house during Hitler's rule, so as to have an excuse for not giving the Nazi salute in greeting.

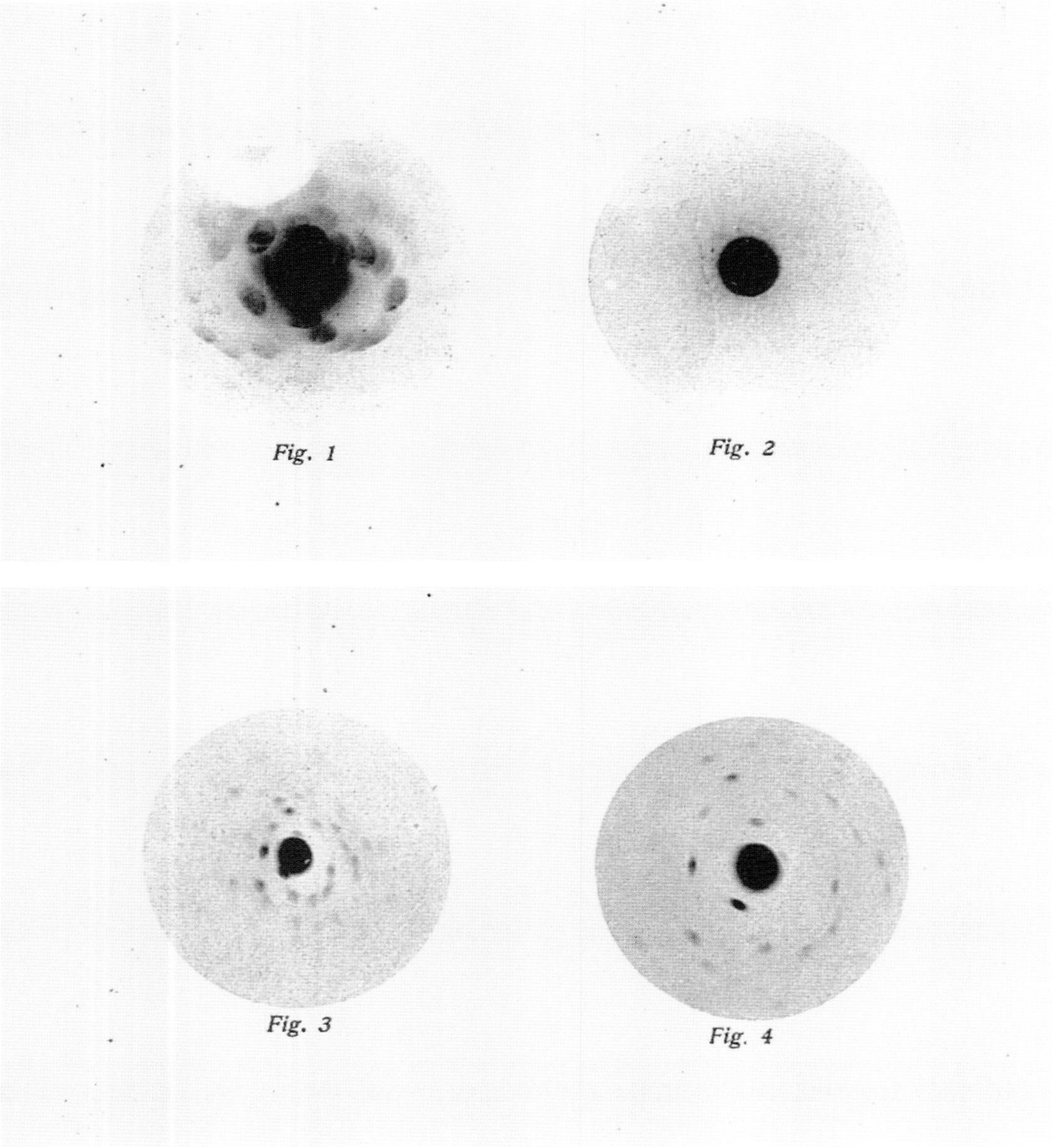

X-ray diffraction patterns of copper vitriol crystals. Walter Friedrich, Paul Knipping, and Max von Laue's *Interferenz-Erscheinungen bei Röntgenstrahlen*, Munich: Sitzungsberichte, 1912, Plate 1, Figs 1–4, Bavarian Academy of Sciences and Humanities, Munich.

X-rays had darkened the emulsion. Those regions of intense X-rays were presumably caused by positive interference of wavelike rays. But Laue, stuck on the idea that the X-rays were re-emitted rather than scattered, didn't at first understand what was going on. Neither did William Bragg,

WALTER FRIEDRICH
1883–1968

As a mere postdoctoral student when he performed his experiment demonstrating X-ray diffraction, Walter Friedrich was omitted from the 1914 Nobel prize awarded to von Laue for the discovery. He subsequently switched to work on medical radiography with X-rays, and after the Second World War he became a citizen of the German Democratic Republic. His consequent isolation from Western scientists, as well as his modesty, probably denied Friedrich the recognition he was due for his work.

See also: Experiment 24, The nature of alpha particles and discovery of the atomic nucleus, 1908–1909 (page 108); Experiment 38, The demonstration of wave-particle duality by electron diffraction, 1927 (page 162).

who repeated the experiments in late 1912 with help from his son Lawrence, then still a research student at Cambridge, after returning to England to occupy a post at the University of Leeds. The senior Bragg initially thought that the X-rays were particles that were somehow being channeled along the gaps between atoms. His son delicately presented another view in a talk in November, where he explained that the X-ray patterns were due to diffraction from the layers of atoms in the crystal. Lawrence Bragg showed how, from the wavelength of the X-rays and the angle of the diffracted spots, one could work out the spacing of the atomic layers, and thus the structure of the crystal. In late 1912 Lawrence reported a similar X-ray diffraction experiment on sheets of the crystalline mineral mica, and in the summer of 1913 the Braggs used the method to work out the crystal structure of diamond.

By the late part of that year the picture was finally becoming clear: X-rays were waves, and their diffraction from crystals supplied a method for deducing crystal structure. This was potentially revolutionary for studying the nature of matter, and that became the topic of the prestigious Solvay Conference on physics in October, where William Bragg presented his results. By 1914 it was clear that a Nobel prize was in prospect, and it was awarded solely to Laue, who had had the idea that sparked it off. The Braggs shared the prize the following year.

Einstein considered the experiment conducted by Friedrich and Knipping to be "among the most glorious that physics has seen so far." It was surely one of the most important for understanding the structure of matter. However, it was also a demonstration that, even with important experiments, their basis and interpretation may be only hazily understood at the time. In retrospect, these stories are told very differently. The accounts commonly given by scientists and recorded in their papers, said Forman, are typically "gross misrepresentations of the conceptual situation."

The German scientist Walter Friedrich photographed on September 19, 1962, courtesy of the German Federal Archives, Koblenz, Germany.

The demonstration of wave-particle duality by electron diffraction (1927)

Q Can particles act like waves?

In 1923, eighteen years after Einstein proposed that light can show particle-like properties, French physicist Louis de Broglie suggested that perhaps the reverse might be true, too: particles such as those that make up atoms might behave like waves. It was little more than a hunch based on analogy, and many thought the idea unlikely. But Einstein was ready to give it a chance. "It looks completely crazy," he wrote, "but it's a completely sound idea." In 1925, Austrian physicist Erwin Schrödinger devised an equation that he thought should predict how such wave-particles behave, which became a cornerstone of quantum theory.

How to prove the idea, though? One of the characteristic features of waves is that they display interference. In 1925, German physicist Walter Elsasser wondered whether interference effects of particles might explain the curious results that two American scientists, Clinton Davisson and Charles Kunsman, had reported in 1921 while studying how electrons bounce off a piece of platinum. Working at the Bell Telephone Laboratories in New York, the two scientists had found that the intensity of the reflected electron beam varied rather puzzlingly at different angles of scattering, rising at some angles and falling at others. No one had been able to explain this result at the time. Faced with de Broglie's idea of wave behavior in particles, Elsasser wondered if the electrons were being diffracted from the crystalline metal, just as X-rays were.

Hearing about Elsasser's idea at a meeting in late 1926, Davisson disagreed that it explained his results. But he thought the basic idea sound and believed he had a way to test it properly. It had come about as a happy accident. Five months before Elsasser's paper appeared, Davisson and his junior colleague Lester Germer had been continuing the electron-scattering experiments when their equipment malfunctioned. A bottle of liquid air used to cool the sample—a piece of nickel—exploded, and the hot metal became covered in a layer of oxide. After the researchers tried to rescue the sample by heating it in hydrogen, they found that the pattern of electrons scattered at different angles had completely changed, with abrupt bumps and dips. They figured that the initial target, a mosaic of many tiny nickel crystals at different angles, had been converted into just a few large crystals. In the many-crystal sample, many diffraction patterns at random angles were superimposed into a blur. But now there was less blurring and more structure in the pattern of scattered electrons. Perhaps if they could create just a single crystal, they would see the expected sharp diffraction peaks.

CLINTON DAVISSON
1881–1951

While studying physics at the University of Chicago, Clinton Davisson became a protégé of Robert Millikan. After completing his doctorate at Princeton in 1911, he took wartime employment in 1917 at the Western Electric Company in New York, which later became Bell Laboratories. After his work on electron diffraction in the 1920s, he stayed at the company working on various problems in physics related to applications in electronics.

By 1927 Davisson and Germer had the setup they needed. They directed a stream of electrons emitted from a hot tungsten filament onto a sample of nickel that had been recrystallized by heating to make it a single crystal, cut with a jeweler's saw to expose a flat facet of the crystal. The researchers used a moveable electron detector to measure how many were scattered from the sample at different angles. Their intensity was greatest at exactly those angles for which X-ray diffraction would produce bright spots.

Others had the same idea of testing de Broglie's proposal of "particle waviness" (or quantum wave-particle duality, as it later became known) by looking for diffraction of electrons. One was the physicist George Paget Thomson, working at the University of Aberdeen in Scotland. He and his student Alexander Reid accelerated electron beams to much higher energies than the Bell labs duo, and fired them at thin films of metals such as gold and platinum, recording the pattern of scattered electrons either on photographic film or on a fluorescent screen. Since Thomson's metal foils were patchworks of little crystals, he didn't see sharp spots, but rather, a series of concentric rings—at precisely the positions expected, given the known crystal structures of the metals.

As well as proving the wave nature of quantum particles, electron diffraction complements X-ray diffraction as a means of working out crystal structures. Moreover, since electrons (unlike X-rays) are charged particles, beams can be focused and diverted by electric fields. This became the basis of the electron microscope. Since the wavelengths of electrons in these beams are much shorter than that of light and typically comparable to that of X-rays, electron microscopes can reveal details too fine for light-based microscopes—now down to the level of individual atoms.

GEORGE PAGET THOMSON
1892–1975

There is a pleasing symmetry in the fact that, while the physicist J. J. Thomson proved that "cathode rays" were not really rays like light but rather particles (electrons), his son George Paget Thomson showed that these particles could act like light waves after all, thanks to their quantum-mechanical nature. George studied at Cambridge and worked for a while in the Cavendish Laboratory that his father previously headed, before taking up a post at Aberdeen. He later worked on nuclear physics and chaired the 1940–1941 MAUD committee that established the feasibility of an atomic bomb, galvanizing the Manhattan Project.

See also: Experiment 37, The diffraction of X-rays by crystals, 1912 (page 158); Experiment 39, The quantum double-slit experiment with single electrons, 1974/1989 (page 164).

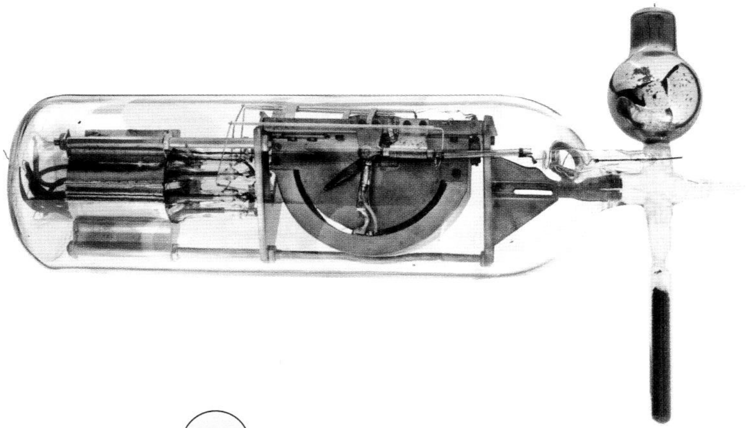

Davisson–Germer electron diffraction apparatus, ca. 1927. This consisted of a glass vessel sealed under high vacuum containing an electron gun that was directed at a nickel target and a movable collector, which detected the resulting scattered elections, National Museum of American History, Washington, DC.

The quantum double-slit experiment with single electrons (1974/1989)

Q Do single quantum particles undergo interference?

Given that wave-like diffraction with beams of electrons was reported by Davisson and Germer in the 1920s, it took a surprisingly long time for someone to recapitulate Thomas Young's double-slit experiment with electrons. The first to do so was Claus Jönsson, a physicist working at the University of Tübingen in Germany, in 1961. To create the double slits, Jönsson used a biprism interferometer: a fine metal filament that is positively electrically charged relative to two metal plates on either side. When an electron beam is passed between the plates, the particles pass on one side of the filament or the other and interfere on the far side. The interference pattern was displayed as a series of bright and dark bands on a fluorescent screen. The strangeness of this wavy nature of quantum particles becomes clear if we imagine dimming the beam until only one particle passes through the slit at a time. In that case we might expect interference to disappear, because the particles must each go through either one slit or the other. The theory of quantum mechanics says otherwise; even if particles traverse the apparatus one at a time, they will still generate an interference pattern so long as two slits—two alternative pathways—are available to them. It is as if a particle passes through both slits at the same time and interferes with itself.

Technically speaking, these single particles are in a state of quantum superposition—a kind of mixture of the two trajectories. It is easy to write down equations for such superpositions, but it is harder to say what the mathematics means physically. In fact, quantum mechanics doesn't permit us to say anything about what the particles are "really doing" before we measure their positions by detecting them on the screen.

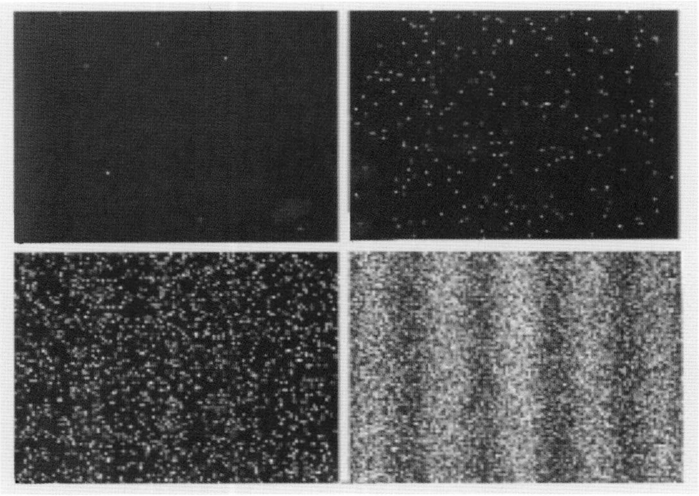

Interference fringes emerge gradually as individual electrons are detected in the quantum double-slit experiment performed by Akira Tonomura and colleagues in 1989.

The possibility of superpositions is what distinguishes quantum mechanics from classical mechanics. In the words of physicist Richard Feynman in the early 1960s, the quantum double-slit experiment with single particles "has in it the heart of quantum mechanics. In reality, it contains the only mystery."

Would electrons really behave this way if they pass through double slits one at a time? Feynman thought it would be an impossible experiment to perform—he was unaware that Jönsson's work had already started to prove otherwise. But Jönsson used streams of many electrons; could it ever really work with single ones?

In 1974, Italian physicists Pier Giorgio Merli, Giulio Pozzi, and GianFranco Missiroli at the University of Bologna demonstrated that it could. Although quantum theory predicts interference for just a single electron, we can't see that because all we observe is a lone particle in the detector. It's only in the cumulative effect of many detections that the interference fringes are revealed. The Italian team inserted an electron biprism into the beam of a standard electron microscope and saw the electrons arriving in their detector system gradually build up into interference fringes.

The Italian team believed they were simply confirming what everyone already knew about quantum wave-particle behavior. They published the work two years later in a paper in the *American Journal of Physics*, which reports mostly on educational and cultural aspects of physics rather than important new discoveries: they thought their results would be useful for demonstrating the wave behavior of electrons to students.

In 1989, Akira Tonomura and his coworkers at the Hitachi Advanced Research Laboratory in Saitama, Japan, showed the same result in more fine-grained detail. They created an electron beam by using a strong electric field to pull electrons from the surface of a hot metal tip—a technique called field emission. To ensure that only one electron passed through the slits (again, a biprism interferometer) at a time, they kept the rate of emission to fewer than 1,000 electrons per second. Their detector could essentially register the arrival of every single particle. At first, this succession of dots looked like a random scattering on the screen. But as more and more arrived, a pattern started to become clear: there were more dots on average in some places than others, in the positions corresponding to the interference fringes. Whereas the particles in the Italian work created rather splotchy signals in the detector that gradually merged into continuous bright fringes, Tonomura's interference patterns emerged like constellations of stars in the gathering dusk.

This experiment demonstrates before our very eyes the profound reality of quantum mechanics. It predicts that individual outcomes of measurements on a quantum system will be random, but that on average the outcomes are governed by predictable probabilities—in this case, probabilities of where electrons will hit the screen—that can be confirmed from many measurements.

AKIRA TONOMURA | 1942–2012

Akira Tonomura's work exemplifies how, in the late twentieth century, a great deal of cutting-edge science was carried out in the laboratories of industrial corporations. After graduating in physics from the University of Tokyo, Tonomura joined Hitachi where he helped develop an electron microscope that uses the principle of holography, recording three-dimensional information in a two-dimensional image. While at the company, he also studied superconductivity and experimentally proved a strange prediction of quantum mechanics in which particles seem to respond to electromagnetic fields even when confined to regions where the fields are zero.

See also: Experiment 34, The wave nature of light, 1802 (page 150); Experiment 38, The demonstration of wave-particle duality by electron diffraction, 1927 (page 162).

Slowing and stopping light (1998-2000)

Q How slowly can light be made to travel?

Although Einstein's theory of special relativity is often said to decree that the speed of light is constant and unvarying, strictly speaking this applies only to the speed in a vacuum. Light *does* change its speed when it travels through another medium, and in fact this change gives rise to refraction—the bending of light beams when they enter a transparent medium like glass or water. The ratio of the light speed in some transparent medium relative to that in a vacuum is proportional to the substance's refractive index.

Normal substances have very modest refractive indices, so the speed of light changes very little: light zips through water much as it does through air or empty space. But in the late 1990s, physicists working at the Rowland Institute for Science in Cambridge, Massachusetts, and Harvard University made a medium with an enormous refractive index that could slow down light to move at a crawl: a mere 17 meters per second, slow enough to be outpaced by a person on a bicycle.

This exotic medium is called a Bose-Einstein condensate (BEC). It consists of a cloud of atoms—the researchers, led by Danish physicist Lene Vestergaard Hau, used sodium atoms—cooled very close to absolute zero. In these conditions the atoms' behavior is dominated by quantum effects that are washed out at higher temperatures. The energy states of atoms are fixed by quantum mechanics and take certain values like the rungs on a ladder. When the gas of atoms is warm, the atoms have a range of energies and they occupy many rungs on the energy ladder. When they have almost no heat at all, they can all "condense" into the lowest energy state, becoming a BEC. Technically, this is possible only if the atoms behave as particles called bosons, which can all occupy the same quantum state. A second class of particle, called fermions (of which electrons are examples), are forbidden by quantum mechanics from all crowding into the same quantum energy state: they can't form a BEC.

Since the atoms in a BEC are in the same state, they act "as one"—as if all part of a kind of giant super-atom. Bose-Einstein condensation is responsible for strange quantum effects such as

Lene Hau adjusting a component on the optics table in her laboratory, as a laser beam travels through the apparatus, Rowland Institute, Harvard.

superconductivity (where a current can flow with no electrical resistance) and superfluidity (where a liquid can flow with no viscosity), which generally only manifest at very low temperatures, where the quantum state is not disrupted by heat. The phenomenon was first demonstrated in cold gases in 1995 by researchers in Colorado.

One of the odd properties of a BEC like this is called electromagnetically induced transparency: one light beam can tweak the way the atoms interact with light so that the previously opaque medium lets a second beam pass right through. It does so in an unusual way that can be thought of as a kind of passing along of photons from one atom to the next by each absorbing and then re-emitting the light. In a BEC, the speed of this convoluted light transmission can be controlled, and Hau and her colleagues were able to make it very slow.

The researchers cooled the sodium atoms, held suspended in a magnetic field, to just 435 billionth of a degree Centigrade above absolute zero. They used a technique called laser cooling to encourage the atoms to relinquish nearly all their heat. In the electromagnetically induced transparent state of the gas induced by a laser beam, the refractive index for a second laser passing though the cloud can vary hugely as that laser's wavelength is changed. In a certain narrow range of wavelengths, it was so large that pulses of light passed through the cloud at just 17 m/s.

LENE HAU | B. 1959

Born in Vejle, Denmark, Lene Vestergaard Hau graduated from Aarhus University in 1984 and began working at the Rowland Institute for Science in 1991. Being trained in theoretical physics, initial funding applications to work on Bose-Einstein condensates were rejected on the grounds she would not possess the necessary skills. Among the many awards she received after proving that assessment gravely mistaken was Denmark's Ole Rømer Medal, named for Hau's compatriot who made the first relatively accurate measurement of the speed of light.

See also: Experiment 35, Measuring the speed of light, 1849 (page 152).

While this work, reported in 1999, wowed the press with its ability to put the brakes on light, there was more to come. At the start of 2001 Hau's team announced that they had stopped light altogether in a BEC of sodium atoms. A laser pulse sent into the cloud would be trapped there until a second laser tweaked the gas to let the pulse emerge again. In this way, the ultracold cloud could act as a kind of memory that stores information as light pulses that can then be read out again. The team believed their method might be useful in information technologies, especially those such as quantum computing and quantum cryptography in which the information is encoded in the quantum states of particles such as photons of light. Hau's later work explored ways to transfer quantum information between matter—quantum devices called qubits—and light.

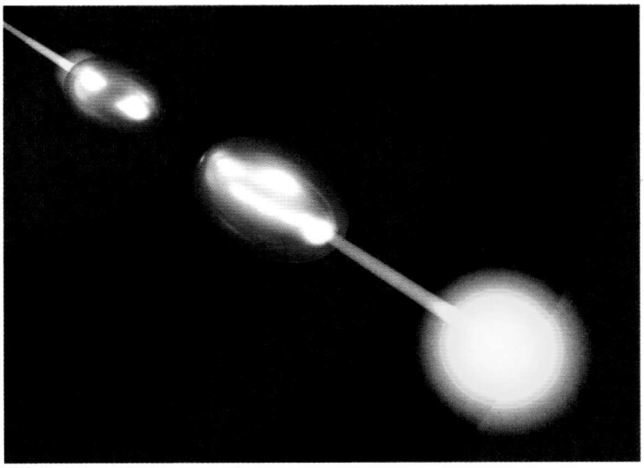

Slowing down light to make it motionless involves transmitting it through an exotic form of matter called a Bose-Einstein condensate, as schematically illustrated here.

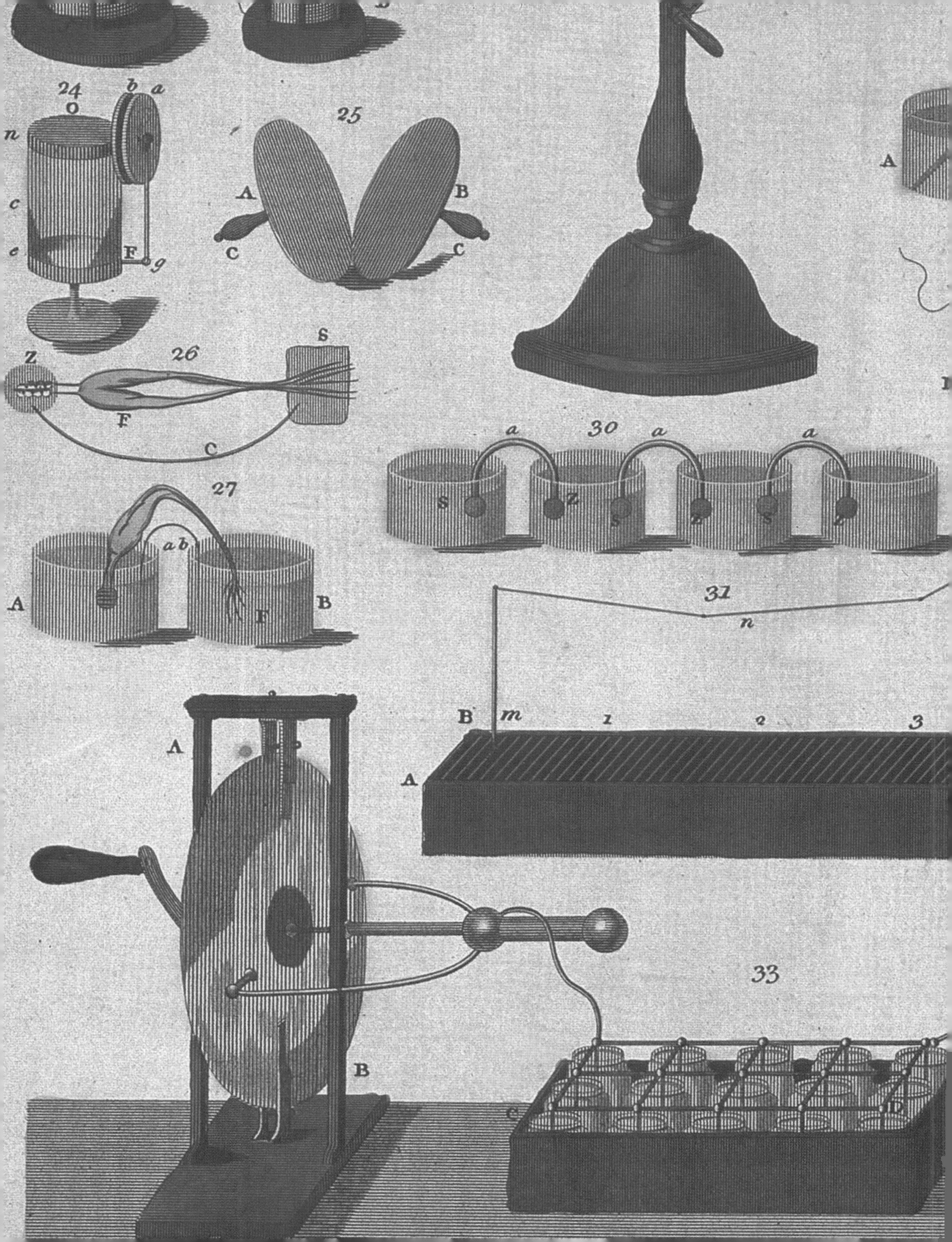
24
b a
o
n
c
e
F g
25
A
B
C
C
26
Z
S
F
C
27
a b
A
B
F
30
a
a
a
S
Z
S
Z
S
Z
31
n
B
m
1
2
3
A
A
B
33
C
D

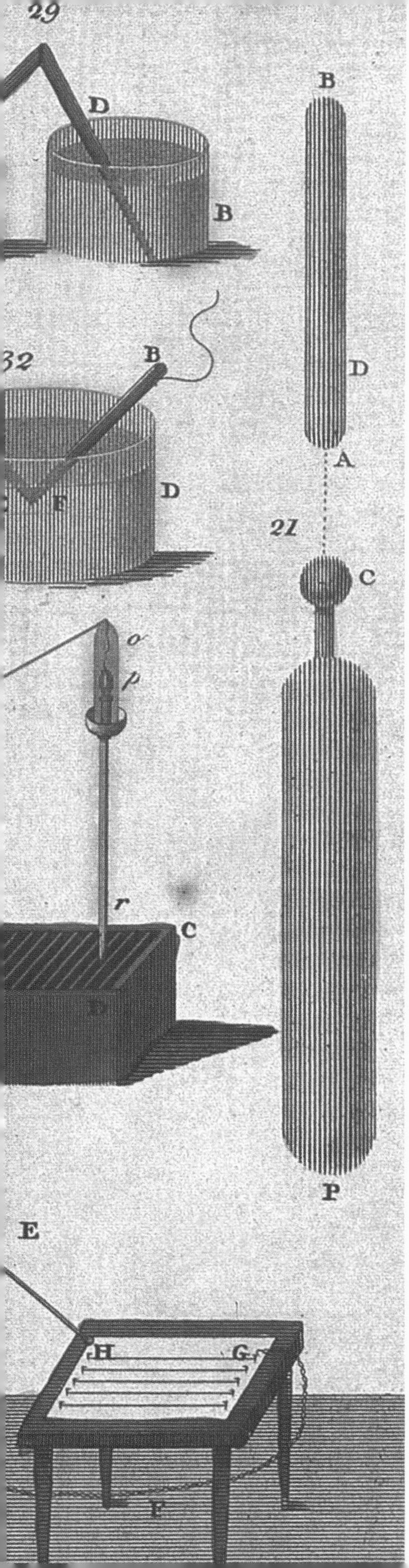

CHAPTER FIVE

What is life?

The microscopic observations of microbes (1670s)

Q Is there life too small to see with the naked eye?

"In every little particle of matter," wrote Robert Hooke in his 1665 treatise on the microscope, *Micrographia*, "we now behold almost as great a variety of Creatures, as we were able before to reckon up in the whole Universe it self." This discovery was as much of a theological as a scientific shock: why had God populated the world in such profuse abandon, even at the tiniest of scales? Hooke and his contemporary microscope enthusiasts, Henry Power in England and Jan Swammerdam in Holland, saw all manner of small creatures, such as water fleas, swimming in drops of water or vinegar, barely visible to the naked eye. Yet there was a bigger surprise to come.

It was perhaps the microscope's capacity to examine the weave of fabrics that attracted the Dutch textiles merchant Antonie van Leeuwenhoek to the instrument. In the 1670s, he began sending his reports of microscopic studies to the Royal Society in London, and in 1676 he reported that he had found tiny creatures—"animalcules"—in a drop of rainwater. These were much smaller than anything Hooke (tasked with verifying the claim for the Royal Society) had seen previously.

The Royal Society was a hub of correspondence between those in Europe who were interested in the new experimental philosophy. The Society's multilingual secretary Henry Oldenburg cultivated a web of contacts, who would send letters detailing their observations. It was a classic "old boys' network" in which the veracity of the claims was judged by reputation—by who had established themselves as a reliable witness. No one at the Royal Society had heard of Leeuwenhoek when Oldenburg received the first of his studies in 1673 (a critique of observations in *Micrographia*). But that report was forwarded by a reliable source, and Oldenburg received an assurance from one of his trusted circle, the Dutch philosopher Constantijn Huygens, that Leeuwenhoek was "of his own nature exceedingly industrious and curious."

Leeuwenhoek was self-taught and made his observations not with one of the beautifully crafted instruments used by Hooke but with his own homemade single-lens microscopes, the

Jan Verkolje's oil on canvas *Portrait of Antonie van Leeuwenhoek, Natural Philosopher and Zoologist in Delft*, 1680–1686, Rijksmuseum, Amsterdam.

ANTONIE VAN LEEUWENHOEK 1632–1723

Aged sixteen, Antonie Leeuwenhoek (he took on the "van" as an affectation later) was apprenticed to a linen-draper in Amsterdam, where he learned to use lenses to "count cloth"—that is, to inspect the density of the weave. In 1654, he married the daughter of a cloth merchant in Delft and became a town official, which gave him the modest independent means to conduct his microscopic research. He continued his observations until the end of his life, still dictating notes on specimens in his final days.

See also: Experiment 44, The role of sperm in fertilization, 1777–1784 (page 180); Experiment 46, The demise of spontaneous generation, 1859 (page 188).

construction of which he kept a jealously guarded secret. These instruments were hard to make and to use, but in principle they could offer greater magnifying power than the double-lens devices.

Leeuwenhoek's letters to Oldenburg describing "animalcules" in fact began in 1674 when he reported his studies of water taken from a Dutch lake. He saw tiny creatures "whereof some were roundish, while others, a bit bigger, consisted of an oval." Some, he said, had little legs and fins, and were "whitish and transparent" or "with green and very glittering little scales." But it was his letter of October 1676 that reported such observations most exhaustively. In samples of rainwater and "pepper water" (to which peppercorns were added) he saw animalcules "so small in my sight, that I judged that even if 100 of these wee animals lay stretched out against one another, they could not reach to the length of a grain of coarse sand." Although some think this indicates Leeuwenhoek was the first to observe bacteria, others believe that his microscopes were only powerful enough to see larger single-celled organisms called protozoa.

Without knowing how Leeuwenhoek made his single-lens microscopes, Hooke had a hard task reproducing these dramatic results. He finally mastered the art of making the little spherical lenses by drawing glass rods into thin strands in a flame and melting the tips into a globule. He wrote with surprise in 1677 that he had "never seen any living creature comparable to these for smallness."

Leeuwenhoek looked at almost anything under his microscope—including his own sperm, rushed decorously from the marital bed. There, too, he found what he called "animalcules" with a long tail that "moved forward with a snake like motion": the first observation of spermatozoa. The Royal Society began sending him lists of substances to study: blood, milk, saliva, sweat, tears, all the excrescences of the human body, and in 1680 they awarded him the rare honor of becoming a Fellow.

A simple silver microscope (about 2 inches tall) with a glass lens made by Antonie van Leeuwenhoek, Rijksmuseum Boerhaave, the Netherlands.

Animal electricity (1780-1790)

Q What is the role of electricity in life?

In the eighteenth century, electricity emerged as one of the most profound and alluring mysteries for the natural philosopher. Electrical phenomena had been recognized for centuries: the ancient Greeks knew that rubbing certain substances, such as *elektron* (amber), made them apt to attract dust and hair, through the effect that we now know to be electrical attraction. This force, like magnetism, seemed to operate through empty space, making it literally "occult": hidden.

It was not until the early 1700s that a command of electricity came within the experimenter's grasp. The English instrument-maker Francis Hauksbee invented a device for producing static electrical charge at will: a glass sphere, resting against a cloth or leather pad, that could be rotated by a handle, in much the same way as you can electrify a balloon by rubbing it against a garment. In the 1720s, the astronomer Stephen Gray used a similar instrument to experiment with electrical conduction. Gray found that electricity could be conducted from the generating device to another object, such as an ivory ball, when a hemp thread connected the two. He thought electricity must be a fluid that passed along the thread. Gray conducted flamboyant experiments in which he electrified young boys from the school connected to his lodgings: suspending them from the ceiling, he drew sparks from the noses of his hapless subjects.

To study this "electrical fluid" more systematically, experimenters needed a way

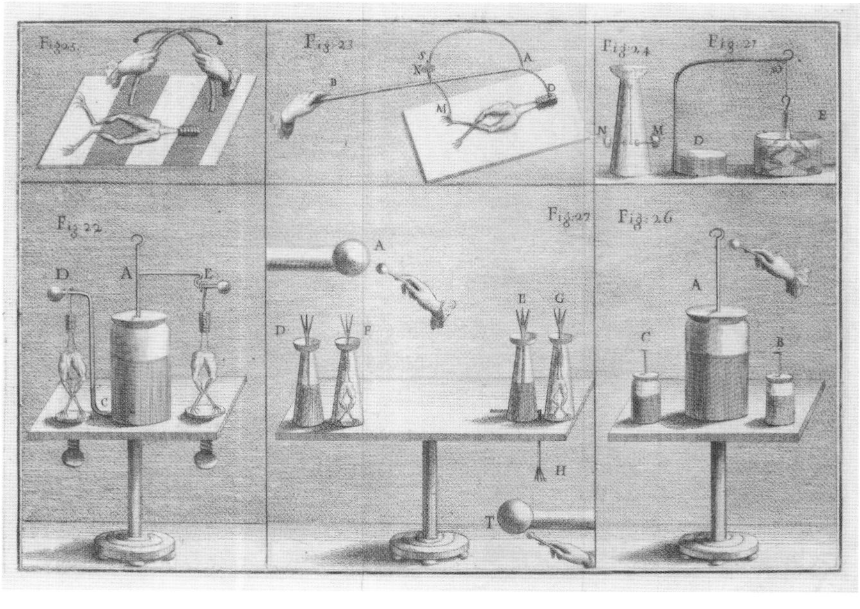

The Italian anatomist Luigi Galvani's experiment that showed how amputated frogs' legs appear to twitch when connected to a source of electricity. Engraving from Galvani's *De viribus electricitatis*, Mutinae: Apud Societatem Typographicam, 1792, Table III, Wellcome Collection, London.

to store it. In 1745 the Polish scientist Ewald von Kleist showed that it could be "bottled" in a glass jar fitted with metal plates. Much the same device was invented independently around the same time by Pieter van Musschenbroek at the University of Leiden, and it became known as the Leyden jar. It was basically a kind of capacitor that accumulated charge on the plates, and could hold enough of it to deliver a nasty shock. When Benjamin Franklin in America heard of Leyden jars in 1746, he began his own experiments and wondered whether the sparks were essentially the same phenomenon as lightning. He proposed the famous kite-flying experiment to try capturing this natural electricity; it is unclear if he ever really carried out that hazardous scheme, although it was popularized by the English scientist Joseph Priestley in his 1767 book *The History and Present State of Electricity.* Franklin suggested that electricity is a fluid that pervades all materials, and that electrical phenomena result from its excess or loss.

The Italian anatomist Luigi Galvani at the University of Bologna became fascinated by electricity in the 1780s. Using Leyden jars, he found that delivering electricity to the bodies of dissected frogs could make them twitch as if with animation. Galvani typically removed the head and upper torso of the frogs, leaving only the legs and the spinal column connected to the source of electricity with wires. He was not the first to notice these electrically stimulated muscle contractions, but his experiments on the phenomenon were much more comprehensive than any attempted previously. He became convinced that animals are themselves generators of what he called "animal electricity," transmitted by the nerves. Perhaps, he said, inside the creatures there is "a kind of circuit of a delicate nerve fluid." In these experiments Galvani was assisted by his wife Lucia Galeazzi and his nephew Giovanni Aldini.

Galvani observed that sometimes contractions might be induced "at a distance," when the frogs' legs were contacted by a metal implement like a knife while an electrical machine produced sparks nearby—even without any direct contact between them. The machines, he concluded, produced their own "electric atmosphere." Was this analogous to atmospheric electricity? In one set of observations, Galvani noticed that dissected frogs' legs that he had left hanging by brass hooks from an iron railing began to twitch during a thunderstorm. He wondered if "such contractions result from atmospheric electricity slowly insinuating itself in the animal." But he found that the movements could also happen on clear days if the corpses of the frogs were pressed against the rail, and he decided that the electricity "was inherent in the animal itself."

Galvani published his observations in *Commentary on the Effect of Electricity on Muscular Motion* (1791). Among its readers was Alessandro Volta, a physicist and chemist (as we would now

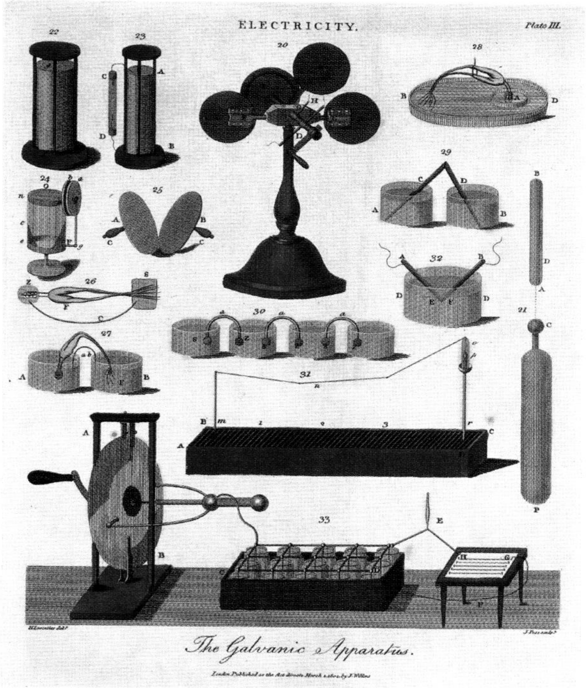

"The Galvanic Apparatus." Colored engraving by J. Pass after H. Lascelles, 1804, Wellcome Collection, London.

LUIGI GALVANI | 1737–1798

Born in Bologna, Italy, Luigi Galvani made his life in that city. He enrolled at the university in 1755 to study medicine and then became a professor of anatomy. He became interested in the growing enthusiasm for the medical uses of electricity—the topic to which he dedicated his life's work. When Napoleon annexed part of northern Italy to create the Cisalpine Republic in 1796, Galvani refused to swear allegiance to the new rulers, was deposed from his position and source of income, and died in poverty.

See also: Experiment 16, The discovery of alkali metals by electrolysis, 1807 (page 72); Experiment 46, The demise of spontaneous generation, 1859 (page 188).

call him) at the University of Pavia. But Volta was unpersuaded by Galvani's interpretations in terms of animal electricity; he believed, on the contrary, that the electricity was not being produced in the animal but in the metals in which it was in contact.

In experiments conducted through the 1790s, Galvani set out to demonstrate that he was correct and Volta's "metal" hypothesis was mistaken. For example, he showed that a frog nerve might be set twitching merely by being brought into contact with a muscle, with no metals involved at all. To each demonstration, Volta found a rejoinder—insisting, for instance, that experimental errors could not be excluded. The controversy

Luigi Galvani's *De viribus electricitatis*, Mutinae: Apud Societatem Typographicam, 1792, Table II, Wellcome Collection, London.

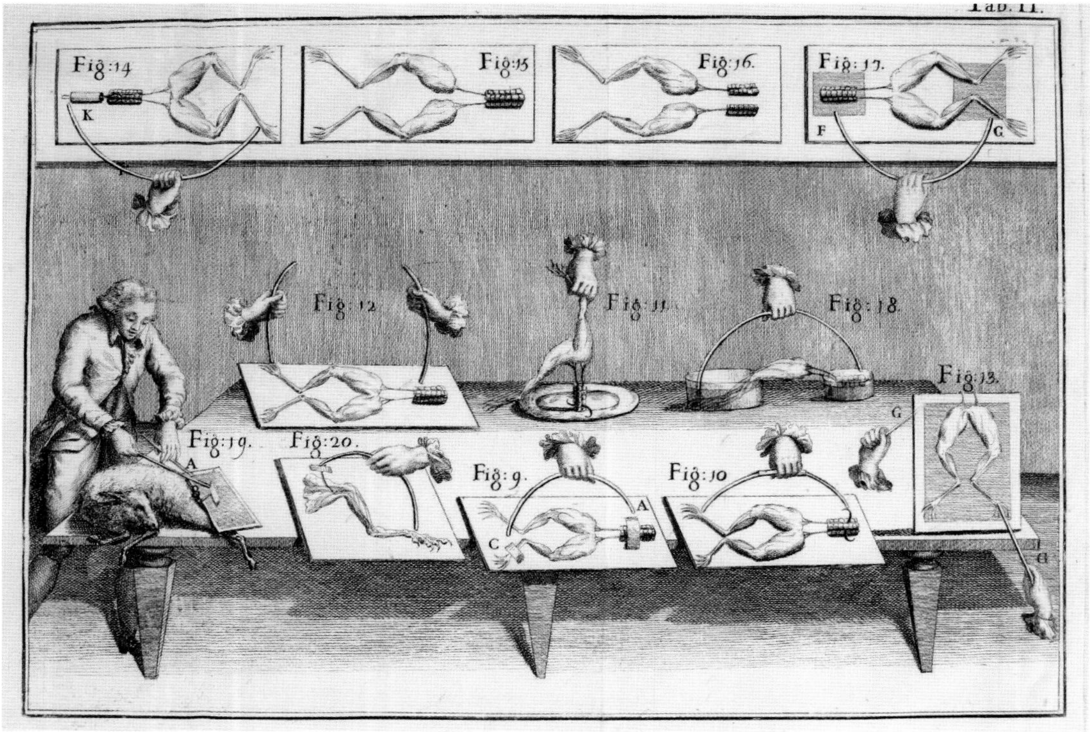

demonstrated how hard it can be—especially in questions concerning living systems—to come up with experiments that offer truly conclusive evidence for or against a hypothesis.

Here both men were correct in a sense. Two dissimilar metals in contact—like brass and iron—do produce an electrical charge that can be drawn off as a current; this was the basis for Volta's "pile," which was the first electrical battery (see page 73). But nerve impulses are indeed intrinsically electrical phenomena, as Galvani insisted.

Galvani's demonstration of a semblance of animation in dead animals seemed to imply some connection between electricity and life. As Emil du Bois-Reymond, the German physiologist who pioneered the understanding of nerve action in the later nineteenth century, put it, thanks to Galvani's experiments "physiologists believed that they grasped in their hands their old dream of a vital force"—or even a kind of elixir of life. Indeed, medical "cures" using galvanism were widely touted for treating paralysis, asthma, constipation, and other conditions.

Aldini advocated for his uncle's belief in the electrical nature of life itself, carrying out dramatic demonstrations of the animating potential of electricity. In 1802 he used a voltaic pile (which he loyally called a galvanic pile) to "reanimate" the severed head of an ox in front of various British dignitaries. A year later, he raised the stakes by using instead the corpse of a criminal recently hanged at London's Newgate prison. As Aldini wrote, "The jaw began to quiver, the adjoining muscles were horribly contorted, and the left eye actually opened." Such a performance, says historian of science Iwan Rhys Morus, "remained susceptible to multiple representations. What Aldini had done could be regarded as anything from an exercise in the possibility of artificial resuscitation, to an effort to resurrect the dead, to a conclusive demonstration of the electrical and material nature of the vital principle."

Illustration from Mary Wollstonecraft Shelley's *Frankenstein; Or, the Modern Prometheus*: London, Henry Colburn and Richard Bentley, 1831.

It is small wonder, then, that when in 1818 Mary Shelley came to describe the reanimation of the body parts assembled by her antihero Victor Frankenstein, she hinted that electricity held the key. Victor recounts how he "infuse[d] a spark of being into the lifeless thing that lay at my feet." In the introduction to her revised 1831 edition of *Frankenstein* she was more explicit as she related the content of the discussions between her husband Percy and Lord Byron that preceded her terrible vision. "Perhaps a corpse would be re-animated," they had said. "Galvanism had given token of such things."

The chemistry of breathing (1775-1790)

Q What are the chemical processes involved in respiration?

In the early sixteenth century, the Swiss physician and alchemist Paracelsus argued that life was basically an (al)chemical process. The processes involved in, say, digestion, are the same as those that an alchemist could conduct in flasks and retorts in the laboratory, albeit orchestrated by a kind of "internal alchemist" that Paracelsus called the archeus.

Although expressed rather fancifully, the idea at the root of these speculations was fruitful and led to the growth of chemistry-based medicine, called iatrochemistry, which flourished in France. While Paracelsus's vision is a long way from that of the great French chemist Antoine Lavoisier in the late eighteenth century, both shared the view that what we would now call biological processes can be understood as chemical reactions, and that correspondences exist between purely chemical transformations and those that govern life.

The late eighteenth century was the era of "pneumatic chemistry," when chemists were preoccupied with understanding "airs": the catch-all term for gases of various kinds. It was in this context that Lavoisier set out to decode the phenomenon of respiration.

When he began these studies in the mid-1770s, Lavoisier was strongly influenced by the work of the English chemist Joseph Priestley, who suggested in his *Experiments and Observations on Different Kinds of Air* (1774–1775) that the body is nourished by the "principle of fire" called phlogiston. This elusive substance was considered to be an element abundant in inflammable substances, which was released when they were burned (see page 70). Combustion ceases when the surrounding air has become saturated in phlogiston and can take no more.

Because the phlogiston theory of combustion is more or less a mirror-image of the true picture involving oxygen (which burning substances take up from the air), it can be confusing to discuss Lavoisier's early experiments on combustion in this language. Priestley had shown that it is possible to make "dephlogisticated air"—what we'd now call oxygen—and that this supports particularly vigorous combustion. Priestley attested that breathing this stuff left him feeling especially invigorated.

Lavoisier's first experimental program on respiration began in 1776, when he made careful measurements of how ordinary (common) air is altered when animals breathe it. He experimented with birds confined in sealed bell jars to see how their respiration changed the trapped air. In a manuscript of October/November 1776, he records that a sparrow died in just under an hour "with a kind of convulsive movement," but that the air in the jar was only slightly decreased in volume. However, its composition had changed: another bird placed in the same atmosphere died almost at once. And the respired air turned lime water cloudy when passed through it: the classic test for what we now call carbon dioxide, which was then known as "fixed air," the term given to it by the Scottish chemist Joseph Black.

Lavoisier and his collaborators made careful measurement of the amounts of fixed and remaining "mephitic" (that is, not life-sustaining) air. He concluded that "respiration acts only on a portion of the air that is respirable, and … this portion does not exceed a quarter of the air of the atmosphere." He compared these results with what happened during ordinary combustion and

Pen and wash drawing attributed to Madame Lavoisier. A man seated in a barrel with his head under a glass canopy has his pulse taken as he breathes, while Antoine Lavoisier dictates to his wife as she writes a report, ca. 1790, Wellcome Collection, London.

with the reactions of metals such as mercury when heated in air. The latter form what were then called calxes—in today's terms, the metals were converted to their oxides by reaction with oxygen. Lavoisier established that the portion of "respirable air" removed in the lungs by breathing is the same as that removed by the formation of a calx such as mercuric oxide. He figured also that the process by which blood circulating from the lungs is turned bright red involves the combination of the blood with this respirable part of air.

This left several questions hanging. Where does the fixed air come from? Is the respirable air (that is, oxygen) converted to it, or merely replaced by fixed air somehow already present in the lungs? Fixed air was also produced by burning charcoal—so is respiration merely a kind of combustion? By 1777 Lavoisier was seeking to unite the chemistry of respiration with that other crucial aspect of combustion: the generation of heat. Perhaps respiration accounts for the warmth of animal bodies? In the early 1780s, he collaborated with the French scientist Pierre-Simon Laplace in experiments that sought to measure the relationship between the heat

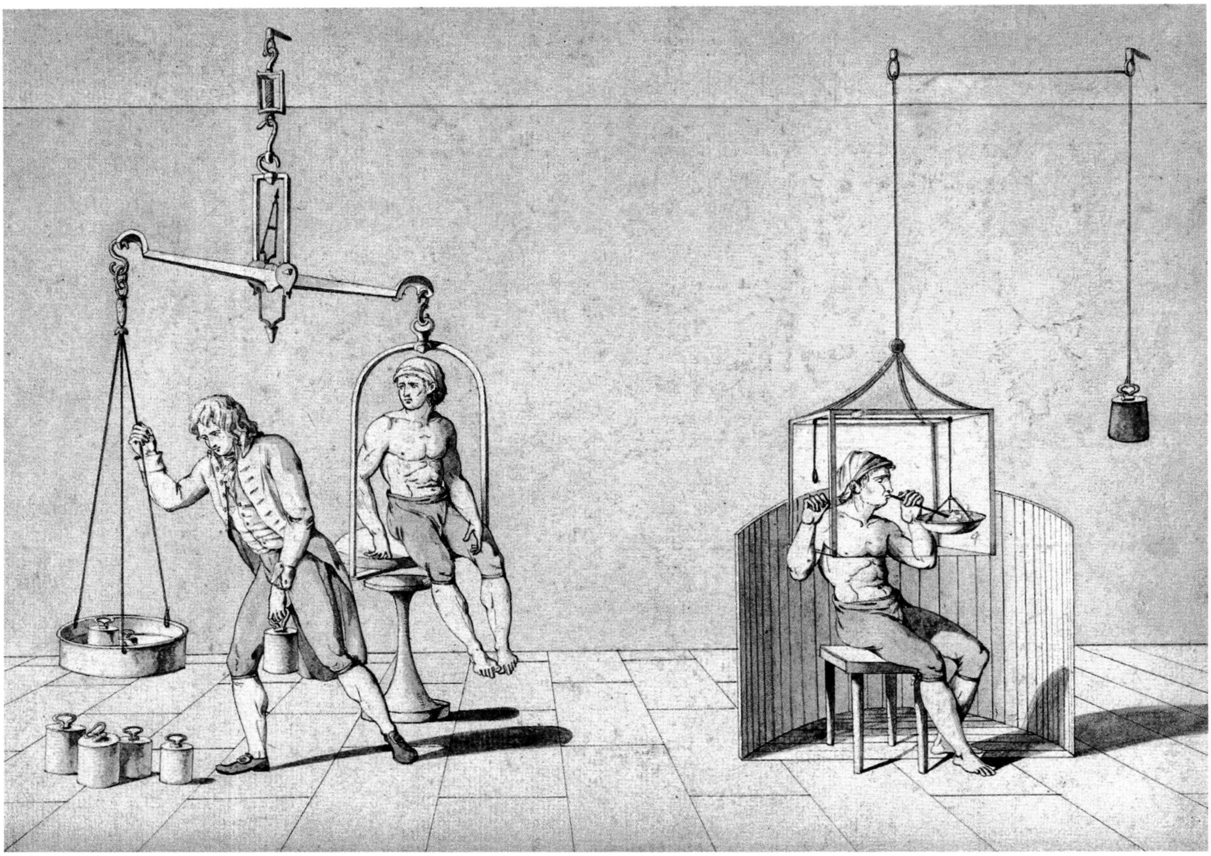

Pen and wash drawing attributed to Madame Lavoisier. A man is weighed on a set of scales and another has his head in a glass container during one of Antoine Lavoisier's experiments with respiration, ca. 1790, Wellcome Collection, London.

animals produce and the amount of fixed air they exhale. Most of these studies used guinea pigs—but as the popular phrase now implies, where experiments on those creatures lead, studies of humans are apt to follow.

Lavoisier did not begin investigating human respiration until the 1790s, at a time when he was very busy with his administrative duties for the French government, for which he was (among other things) a tax collector. Some of his experiments would have been conducted by his assistants, in particular a young protégé named Armand Seguin. Lavoisier was also assisted in the laboratory by his wife Marie-Anne Pierette Paulze Lavoisier, who was deeply knowledgeable about the debates in pneumatic chemistry. Because Antoine spoke and read very little English, Marie-Anne translated for him the works of Priestley, Black, and others, and she also kept laboratory records, including sketches of the experiments that benefitted from her training in drawing by the artist Jacques-Louis David. By today's standards, Mme Lavoisier would be deemed, like Seguin, a coauthor of the work.

Lavoisier's plan was to measure the respiratory processes of humans with the same precision he had used for animals. Obviously, he could not place a person in a bell jar until they expired, but instead he made masks that covered the face so that all the gases inhaled and exhaled could be quantified. To be the subject of such a study was not a pleasant experience. "As painful, disagreeable, and even dangerous as these experiments to which one had to submit oneself are," Lavoisier wrote, "M. Seguin wished them to be carried out on himself."

By this stage Lavoisier had ditched phlogiston in favor of his own *oxygène* theory of combustion. He now regarded common air as a mixture of oxygen and "azotic" (meaning not life-supporting) gas: this *azote* (the name still used in France) is what we call nitrogen. In late 1790, Lavoisier wrote to Black to say that he had measured the volume of oxygen gas that "a man at rest and in abstinence [of food] consumes," and how this changes with exercise and room temperature. He concluded that azotic gas "does not serve any purpose in the act of respiration, and it leaves the lungs in the same quantity and condition as it entered."

During breathing, Lavoisier concluded, oxygen extracts from the blood circulating in the lungs "a portion of carbon," producing fixed air. At the same time, hydrogen in the blood combines with oxygen to make water, which is expelled as moisture. In return, as it were, air deposits in the blood "a portion of caloric"—that is, heat, which Lavoisier regarded as a kind of fluid substance—that "is distributed within the blood, by the circulation, into all parts of the animal economy." Whether the oxygen is converted to fixed air (as in charcoal burning) or is just replaced by it, the experiments did not distinguish. Lavoisier knew the limits of what he could infer with confidence.

Lavoisier claimed even to be able to quantify how much oxygen was needed and how much heat generated, "in the work of a philosopher who reflects [and] of a scholar who writes." He was on the verge of launching the field of bioenergetics: the conversion of chemical to mechanical energy in the body via metabolic reactions. Although not confined to a single experiment—the question was much too complicated to be answered that way—Lavoisier's campaign over fifteen years or so to understand respiration laid the foundations of metabolic biochemistry and was a masterpiece of careful observation and quantification coupled with theorizing that was at the same time both bold and cautious.

Evidently, Lavoisier realized, the oxygen consumed, and the food burned up as a result, depends on the strenuousness of the body's toil. "The poor man who lives by the work of his hands" thus needs more sustenance than the idle, wealthy man. This hint of the economic disparities of eighteenth-century life looks in retrospect like a grim harbinger of the revolution that was about to consume France, and extinguish its great savant Antoine Lavoisier (see page 71).

MARIE-ANNE PIERETTE PAULZE LAVOISIER | 1758–1836

In 1771 the French businessman Jacques Paulze married his 13-year-old daughter Marie-Anne to a brilliant young associate in his tax company, Antoine Lavoisier, to avoid her having to marry the much older brother of a baroness. It proved a good match: not only did they like each other but Marie-Anne was also smart and perceptive. Lavoisier's friends were soon calling her his "philosophical wife": she taught herself chemistry, kept notes of his results, drew his illustrations, translated English papers, and organized and hosted philosophical soirées with eminent guests from France and abroad. When Lavoisier was arrested during the French Revolution in 1793, she petitioned for his release and was briefly imprisoned herself. After his execution she left for England and married the physicist Benjamin Thompson (Count Rumford). The union was not a happy one, and she eventually returned to France, dying in Paris at the age of 78.

See also: Experiment 15, The discovery of oxygen, 1774–1780s (page 70).

The role of sperm in fertilization (1777-1784)

Q What induces eggs to develop?

It was always clear that for the reproduction of humans and other mammals, both the male and the female have a role. But what role, exactly? Aristotle proposed that both sexes contribute a kind of generative principle called *sperma*, which combine to produce the rational human soul in the growing fetus in a process he called epigenesis. Reflecting the chauvinistic attitude that shaped most theories of procreation until modern times, the male principle was considered the active element, which grew like a seed in the passive receptacle supplied by the woman. In one of the earliest recorded examples of experimental science, Aristotle carefully opened and examined chick eggs at different stages from fertilization to birth in order to watch the development of the fetuses.

In the seventeenth century, the English physician William Harvey, while largely endorsing the Aristotelian position, placed more emphasis on the role of the egg from the female. *Ex ovo omnia*, as he put it in 1651: everything comes from an egg, a position called ovism. But Antonie van Leeuwenhoek's microscopic observations of spermatozoa in the 1670s (see page 171) led to the notion that the developing body is somehow inherent already in the head of the worm-like entities seen in sperm (spermatozoa literally means "sperm animals"). The concept was strikingly illustrated in 1694 by the Dutch microscopist Nicolaas Hartsoeker, who drew a sperm with a fetal homunculus packed into the head, complete with tiny limbs. In this *preformationist* view the body was already fully formed, whereas in the epigenetic view it developed from an unstructured seed.

It was all largely conjecture, because of the difficulties of making observations and experiments on human conception. Then, as still today, much of what was known about embryology relied on studies of other animals. In the mid-eighteenth century, an Italian physiologist and priest named Lazzaro Spallanzani set out to investigate the precise role of male semen by studying reproduction in frogs.

Spallanzani has been described as having a "lust for knowledge": a passion that sometimes seemed to exceed propriety, as when he was said to have begun expounding enthusiastically to

LAZZARO SPALLANZANI
1729–1799

Like several Renaissance and Enlightenment natural philosophers, Lazzaro Spallanzani was also a cleric, ordained as a Catholic priest. As well as his work on reproduction, he also conducted important experiments on spontaneous generation, showing that liquids boiled to kill off all microorganisms would not then generate them if kept tightly sealed. He also worked on circulation and respiration, fossils, and bat echolocation. He was elected a foreign Fellow of the Royal Society in 1775.

See also: Experiment 41, The microscopic observations of microbes, 1670s (page 170); Experiment 46, The demise of spontaneous generation, 1859 (page 188).

a group of dignitaries about the mating of frogs he had noticed in a body of water during travels in Constantinople. That was surely a rather indecent topic of discourse for a man who had been ordained in the Church.

Frogs do not actually copulate, though. Rather, the female lays her eggs, onto which the male then deposits his semen. Although Spallanzani shared Harvey's epigenetic view of development from a fertilized egg, he suspected that the spermatozoa play no role, but are instead a kind of parasite. It is the thinner, liquid component of the semen, Spallanzani thought, that provides the generative principle.

To test that idea, Spallanzani needed to collect frog sperm and separate the microscopic "worms" from the seminal fluid. He borrowed an idea from the French scientist René Antoine de Réaumur, who in 1736 attempted to study fertilization in frogs by fitting the males with tiny trousers made from taffeta and pig's bladder, so that he might collect and study their sperm. He had little success with this, however, because the frogs wriggled out of their garments. But Spallanzani's prophylactic frog trousers were more successful. By applying some of the semen obtained in this way to frog eggs, he conducted the first known example of artificial insemination.

Spallanzani never managed to elucidate the connection between spermatozoa and fertilization, however. After filtering the collected sperm to separate the liquid from the thicker residue containing spermatozoa, he found that only the latter could cause fertilization—and yet, to the bemusement of many historians of science, he still believed that the fertilizing capacity resided with the liquid. In one set of experiments he kept toad sperm and eggs on pocket-watch glasses just a few millimeters apart to see if the eggs might be fertilized by some intangible "aura" emanating from the sperm (they were not, of course). Some think Spallanzani was too much in thrall to a kind of preformationist ovism, believing that the embryo's form was already dormant in the egg and needed only the barest stimulus from semen to start it developing. At any rate, he exemplifies how, even if a scientist does the right kind of experiment, there's no guarantee that they will draw the correct conclusion from it.

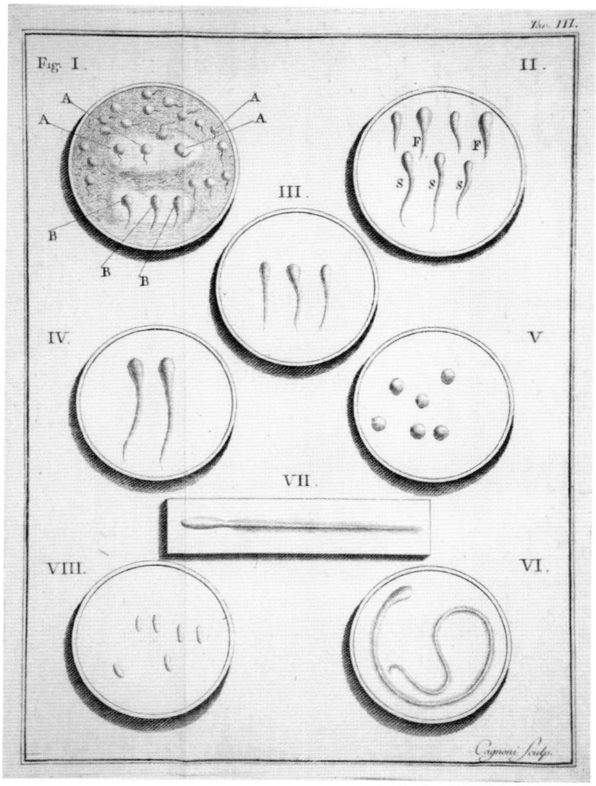

The sperm of various animals. From Lazzaro Spallanzani's *Opuscoli di fisica animale, e vegetabile*, Modena: Presso la Societa Tipografica, 1776, Vol. 2, Plate III, Wellcome Collection, London.

Spallanzani later extended his investigations to dogs, collecting semen from a male and injecting it with a syringe into the uterus of a bitch. Seeing the resulting pregnancy and the birth of pups was, he wrote, "one of the greatest joys of my life." He also experimented with preserving sperm using ice to see if it retained a capacity to fertilize.

Thought experiments

Some "classic" experiments of the past were possibly never done, or not by those who first described them. We saw that Galileo probably never climbed the Tower of Pisa to drop objects (see page 34); similarly, Benjamin Franklin probably never investigated electricity by flying a kite in a thunderstorm (contemporaneous illustrations show this being done from a window). Natural philosophers of earlier times might assume they already knew how an experiment would turn out, saving them the trouble of actually doing it. But some experiments are articulated with the explicit intention that they will not be carried out. They are "what if" exercises of reasoning called thought experiments.

Thought experiments have been used since antiquity and were popular in the Middle Ages—a time, according to historian Edward Grant, when "the imagination became a formidable instrument in natural philosophy." A good example still invoked today in discussions of free will is Buridan's ass, a scenario attributed to the fourteenth-century French scholar Jean Buridan. Imagine, he said, an ass that is both hungry and thirsty, placed midway between a bale of hay and a bucket of water. With nothing to break the deadlock of rational choice, the ass would die in indecision.

Are thought experiments useful?

You might wonder why, if an experiment is indeed meant to test a hypothesis, what function a thought experiment can serve. If you believe you know the answer well enough that it's not worth finding it out, what is the point of the exercise? But thought experiments, in fact, have many uses. They have been invoked to support or illustrate a theory, or conversely to refute it (by showing circumstances in which it would lead to seemingly absurd consequences). They can be a stimulus to thought: the precursor to theorizing. That was how Albert Einstein tended to use them. He claimed to have been led to his theory of special relativity in 1905 by having imagined, aged only sixteen, what it would mean to chase a light beam. And he began working toward his theory of general relativity by imagining someone conducting physics experiments in an enclosed space, like an elevator, that is being accelerated upward, mimicking the effect of gravity. These examples typify the narrative nature of thought experiments: one describes the situation, lets it play out, sees what happens, and draws a conclusion.

Thought experiments can also be used for educational purposes: they might offer an intuitive illustration of some point that otherwise seems obscure. Isaac Newton found such a way to explain his claim that the same gravitational force (as we'd now call it) that makes an object fall holds the Moon in its orbit around the Earth. Why then doesn't the Moon fall too? But it *is* falling, Newton said: forever. He imagined firing a cannonball from a mountaintop with ever greater force, so it descends to the ground at ever greater distances away. Due to the curvature of the Earth, eventually there comes a point where the planet's surface curves away as fast as the ball falls toward it, and so it orbits forever.

Maxwell's demon and Schrödinger's cat

Although Newton's cannonball experiment could be conducted in real life, this isn't possible with all thought experiments. Indeed, one objection is that they may rely too much on our intuition about the outcome. Take James Clerk Maxwell's proposal for how a microscopic, intelligent being (later dubbed a demon) could undermine the Second Law of Thermodynamics by intervening in the motions of atoms. The Second Law says that heat will always flow from hot to cold, so temperature differences

Buridan's ass: a child feeding a donkey. Fifth-century mosaic made from limestone and terracotta, Great Palace Mosaics Museum, Blue Mosque, Istanbul, Turkey.

inexorably dissipate. In 1867 Maxwell argued that the observant demon could separate "hot" (fast-moving) particles in a gas from cold, slower ones, by opening and closing a trapdoor between two gas-filled compartments to isolate hot gas on one side and cold on the other—creating a temperature difference in a gas of initially uniform temperature. The Second Law is considered inviolable, but it took about a hundred years to spot the flaw in Maxwell's reasoning: the need for the demon to clear its memory of particle motions every so often.

On the other hand, while Maxwell imagined his thought experiment to be purely hypothetical, modern technologies have made it possible to conduct real experiments on analogous scenarios in which the motions of microscopic particles are observed and manipulated. It demonstrates a connection between information and energy, whereby information can act as a kind of fuel. The thought experiment did not prove what Maxwell thought, but it has been abundantly fertile.

Thought experiments feature prominently in the development of quantum theory, the most famous being Schrödinger's cat, proposed in 1935 by Erwin Schrödinger. This was one of the "refuting" varieties: Schrödinger argued that it exposed the absurdity of the idea that an observation is required to transform the probabilities of experimental outcomes supplied by quantum mechanics into actual realities. By linking the fate of a cat hidden in a box to the outcome of a single quantum event, he argued that such an interpretation implied the impossible situation of a cat that was, until observed, both alive and dead at the same time.

Despite its popularity, Schrödinger's cat is rather dubious from a physics perspective. Still, modern experimental techniques are again bringing the thought experiment close to experimental realization—if not for a cat, then maybe for a virus or microorganism. This illustrates a problem with thought experiments: it can be hard to tell whether they respect physical laws, as they should. Some believe thought experiments shouldn't be too fantastical—like, for example, philosopher Derek Parfit's exploration of selfhood that imagined people who divide like amoebae. Others, like Pierre Duhem, have called for thought experiments to be banned in science. But most scientists continue to see them as a valuable way to play with ideas—and perhaps to motivate real experimental science.

45

The principles of inheritance (1856-1863)

Q What are the biological laws governing heredity?

The principles of inheritance have been pondered since the start of civilization—by natural philosophers, by farmers and animal breeders, and by monarchs and aristocrats eager to preserve the supposed superiority of their lineage. It was obvious that offspring can resemble their parents in looks and other traits—a fact exploited in livestock breeding long before science had much to say about the matter. The nineteenth-century friar Gregor Mendel, working at St. Thomas's Abbey in Brno, Moravia, was one of the first to look quantitatively at the issue. He was not really investigating heredity itself, but rather the question of hybridization: what happens when individuals with different traits are crossbred.

Starting in the 1850s, Mendel studied the issue by crossing plants, specifically edible peas (*Pisum sativum*). Although pea plants are "true breeding"—able to fertilize (pollinate) themselves—Mendel wondered what happened when those with different characteristics are crossbred, letting one plant pollinate another. There are various distinguishing features the plants can have; Mendel looked in particular at those that produced peas of different color (yellow or green) and texture (smooth or wrinkled), but also at the color and shape of the pods, the position of the flowers, and other features.

For some of these traits, Mendel found that when he crossed plants with two different forms of the trait—one with yellow peas and one with green, say—the traits reappeared in the offspring essentially unchanged (rather than blended to yellow-green), but in rather fixed ratios, typically 3:1. He called the character that appeared in the highest proportion "dominant." Sometimes a trait would vanish in the offspring only to reappear in the next generation. He called these "recessive."

Mendel realized that these results could be explained if we assume that each plant carries two trait-determining factors. They could both imprint the dominant trait (*dd*), or both be recessive (*rr*), or there could be one of each (*dr*). If a plant has a dominant and recessive factor, only the former is manifested in the visible trait; only if it has just two copies of *r* does that characteristic become manifest. A ratio of 3:1 in the dominant and recessive traits in offspring is then just what we'd

GREGOR MENDEL | 1822–1884

Gregor Mendel's homeland on the Moravian–Silesian border was part of the Austro-Hungarian Empire in the early nineteenth century. He entered the monastery of St. Thomas at Brno (Brünn) in 1843; scientific studies of breeding were already by then an established aspect of the institution. Although he also studied physics and mathematics at Vienna while completing his training as a priest, he had to cease all scientific work after he was elected abbot in 1868. When he died in 1884, his successor burned all his papers.

See also: Experiment 48, The random nature of genetic mutations, 1943 (page 194); Experiment 49, The proof that DNA is the genetic material, 1951–1952 (page 197).

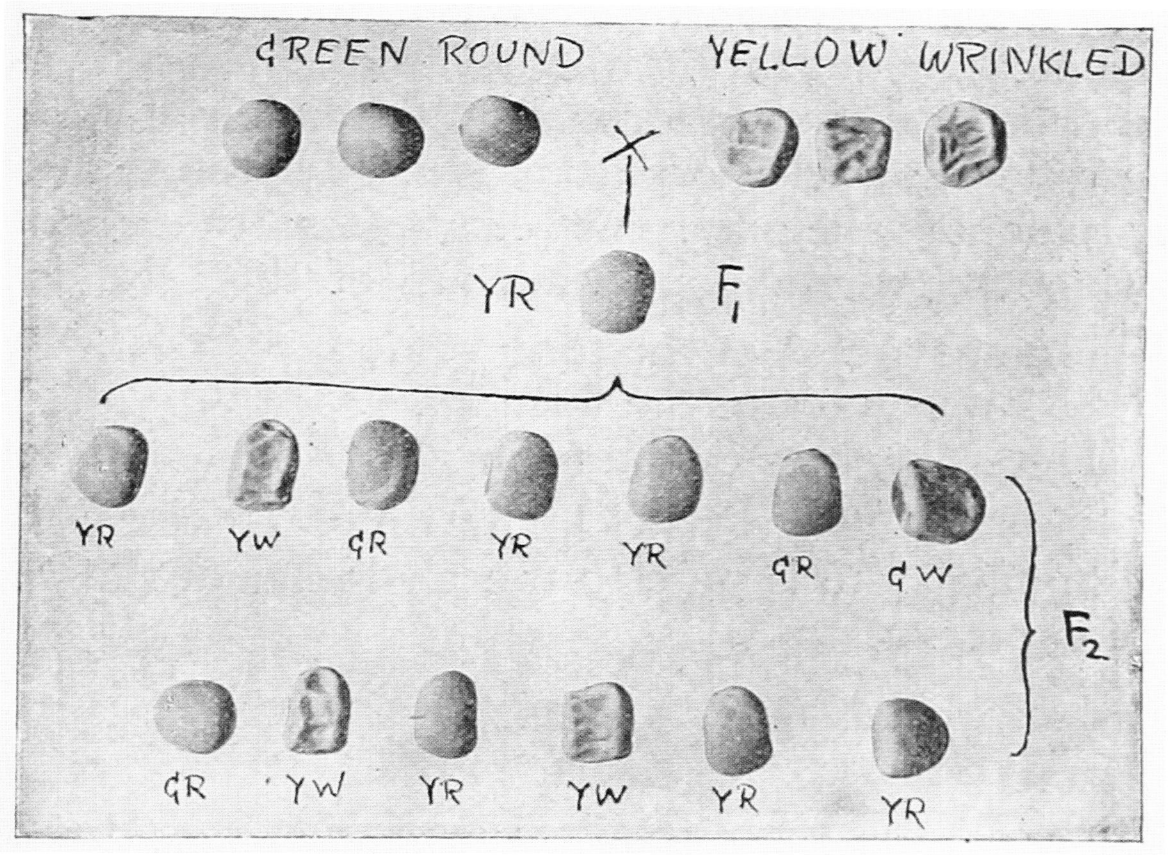

Inheritance of seed characters in peas. From William Bateson's *Mendel's Principles of Heredity*, Cambridge: Cambridge University Press, 1913, Fig. 3, Gerstein Science Information Centre, University of Toronto.

expect from crossed plants that themselves have a dominant (*d*) and recessive (*r*) factor, assuming that these factors are distributed randomly in the offspring. If all the offspring happen to turn out as *dr* or *dd*, the recessive character won't be evident among them at all—but it could nonetheless reappear in a subsequent generation if two *dr* plants are crossed to produce an *rr* variant.

Mendel's basic observations of hybridization were not entirely new, but he was the first to spot the patterns and explain them with dominant and recessive factors. Today we understand these as two versions (called alleles) of a gene that influences the trait in question. However, for Mendel, this "law of hybridization" was simply a description of his observations, to which he attached no particular physical mechanism.

Mendel concluded his experiments in 1863 and presented them in a paper of 1865 to the Brno Natural Science Society. His work was hardly cited at all by other scientists until the end of the nineteenth century. It is often said to have been "forgotten," but it might be better to say that other researchers did not find it terribly relevant or groundbreaking. Some considered that it showed only what was already widely known: that the offspring of parents with different traits, rather than blending those traits over successive

Gregor Mendel, Abbot of Brno. From William Bateson's *Mendel's Principles of Heredity: A Defence*, Cambridge: Cambridge University Press, 1902, Wellcome Collection, London.

generations, tend instead to revert to the traits found in the parents.

That issue touched on what seemed to many naturalists to be the key question about inheritance: are parental traits blended or do they stay distinct? Mendel's experiments with peas seemed to show the latter, but Charles Darwin's theory of natural selection, unveiled in his 1859 book *On the Origin of Species*, appeared to imply that mixing should occur. The truth is that both are possible—as, for example, in skin pigmentation (graduated) or eye color (discrete) in humans. Traits that are inherited intact but in the various proportions among offspring that Mendel observed are now said to be Mendelian.

Legend has it that Mendel's treatise describing the results of his work sat in Darwin's library at Down House, in Sussex, England, unopened and the pages uncut. It is often suggested that, had Darwin read Mendel's work, it could have demonstrated to him how his theory might work via the passage from parent to offspring of physical entities (now called genes) that determine traits. But in fact there is no firm proof that Darwin *did* possess a copy of Mendel's book. And even if he did, and had bothered to read it, it probably wouldn't have seemed terribly relevant to his evolutionary theory.

Darwin did later propose that inheritance relies on trait-determining "particulate factors"—perhaps molecules—in the parent's sex cells that unite to form their offspring. He called this idea pangenesis, and the factors were called pangenes by the Dutch zoologist Hugo de Vries in 1907—later shortened to "genes." But Darwin's pangenes weren't much like the genes now considered to be encoded in the DNA molecules of chromosomes. For one thing, Darwin considered that they could be altered during an organism's own lifetime by what it experienced. In discussions among de Vries and others of what genes can and can't be, the significance of Mendel's results finally became clear. It wasn't so much that researchers finally understood what Mendel had been saying; rather, the Austrian friar's results finally *acquired* importance because of the new context supplied by the idea of genes. As geneticist Ronald Fisher put it, "Each generation found in Mendel's paper only what it expected to find ... [and] ignored what did not confirm its own expectations."

At first, however, Mendel's experiments seemed to *conflict* with the predictions of natural selection, and so Darwin's theory appeared to flounder at the end of the nineteenth century. Mendel's work seemed to show that inheritance was "discontinuous," like a series of steps, while Darwin's theory supposed it to be gradual, like a ramp. Darwinism was rescued only after population geneticists and statisticians, including Fisher, showed in the 1920s how traits influenced by several different genes could appear in "blended" form. Far from contradicting Darwinian theory, Mendel's pea experiments now appeared to support it. This unification of Darwinism and Mendelism was dubbed the Modern Synthesis by biologist Julian Huxley in 1942, and it formed the conceptual framework for twentieth-century genetics and evolutionary theory.

Photograph of Gregor Mendel's garden at St. Thomas's Abbey at Brno, where he became the abbot in 1868. From the papers of William Bateson, 1910, Department of Archives and Modern Manuscripts, Cambridge University Library.

The demise of spontaneous generation (1859)

Q Can life arise spontaneously in non-living matter?

To the ancient Greeks, animation of matter was a matter of degree. Stones were lifeless, but metals, which were sometimes found to grow in branched forms that looked "organic," had something of the vegetative about them. Plants more so, and animals (especially humans) were the highest forms of life.

If this were so, no insurmountable barrier seemed to separate the animate from the inanimate. It was widely thought in ancient and medieval times that the simplest of animals, such as worms, insects, and even rodents, can spring spontaneously from lifeless matter: a belief that was known as spontaneous generation. Why else did maggots appear in decaying meat, mice throng in stored grain, or frogs creep forth from the mud of the Nile after the annual flood?

The theory of spontaneous generation was challenged in the seventeenth century by the Italian scholar Francesco Redi, who conducted a remarkably modern experiment in order to test it. He left meat to rot in three jars: one jar was left open to the air (which we'd now call the "control" sample), one covered in gauze, and one sealed tightly. It was only in the open jar that maggots appeared in the meat. Redi attributed this result to flies laying their eggs in the decaying matter, which they obviously could not do for the covered and sealed jars.

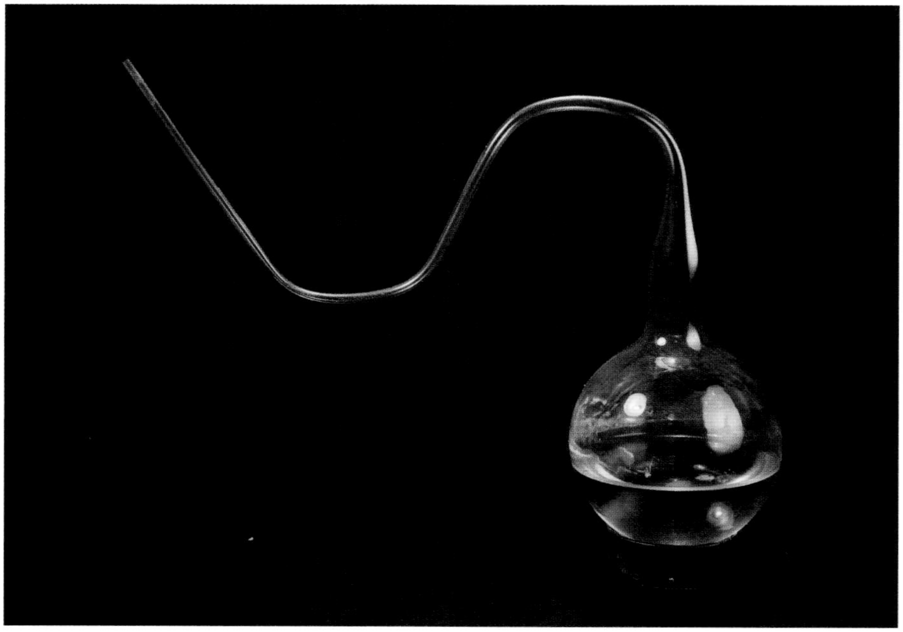

Left: A swan-necked flask from the nineteenth century used by Louis Pasteur in his experiments to disprove the doctrine of spontaneous generation, Wellcome Collection, London.

Opposite: Albert Edelfelt's oil on canvas *Louis Pasteur,* 1885, Musée d'Orsay, Paris.

That didn't settle the matter, however. After the discovery of microorganisms in the late seventeenth century, others argued about whether such microbes could appear spontaneously in broths that were sealed and sterilized by heat to kill off pre-existing microbes. The difficulty of fully sterilizing such a mixture led to ambiguous results.

In the early nineteenth century this debate was entrained with the notion of vitalism: the idea that living things are imbued with a "vital force" that makes them fundamentally distinct from non-living matter. The demonstration in 1828 by the German chemist Friedrich Wöhler that a substance hitherto associated only with living systems—urea—could be synthesized from inorganic chemicals contributed to (but certainly did not bring about, as is sometimes suggested) the demise of vitalism. But it was probably the experiments conducted by the French scientist Louis Pasteur in the late 1850s that did most to bury both the doctrines of spontaneous generation and vitalism.

Pasteur was motivated by a prize offered by the French Academy of Sciences to resolve the first of these issues. Pasteur, today often regarded as the father of microbiology, was an expert in fermentation processes involving microbes, such as those which caused milk and wine to spoil. (He was employed for a time as a consultant for the French cheese industry.) Pasteur knew that microbes such as bacteria are less easily excluded from decaying foods than are maggots: in 1858 he showed that they can be carried in air, being collected in abundance in a filter of gun-cotton.

If there exists a vital force that can animate inert matter, one might expect it to be even more tenuous and pervasive than airborne microbes. So Pasteur devised an experiment that would, in his view, keep out microbes from a meat broth but enable any putative vital force access to it. He had a glassmaker produce a set of elegant, delicate flasks with long, narrow tubes at the neck, bent into an S-shape that was referred to as a "swan neck." Pasteur's hypothesis was that no airborne microorganism would be able to negotiate this passage without getting stuck on the glass—but the air itself still could. So if spontaneous generation due to a vital force were possible, broth in the flask that was first well boiled to thoroughly sterilize it would eventually turn cloudy with decay. But if all decay and fermentation depended on the ingress of microbes—"germs," as Pasteur called them—then the broth would remain sterile.

The latter was what he observed. If the swan neck was broken off, meanwhile, the broth would indeed spoil. There was no spontaneous generation, but only contamination. For this work Pasteur was awarded the Alhumbert Prize by the Academy of Sciences in 1862. As his son-in-law René Vallery-Radot reported in his biography (which was rather hagiographic and must be taken with a pinch of salt), Pasteur declared that "Never will the doctrine of spontaneous generation recover from the mortal blow of this simple experiment." Whether or not we can believe that account, the sentiment was right: as Pasteur said in an 1864 lecture, *Omne vivum ex vivo*: all living things arise from another living thing.

FRANCESCO REDI | 1626–1697

The seventeenth century saw the rise of experimental science as a means of challenging old beliefs through direct observation. Francesco Redi, an Italian physician and naturalist, was one of the rationalistic sceptics of the age who confronted such ideas with empirical tests. As well as disproving (in his view) the notion of spontaneous generation, he demonstrated errors in several beliefs about poisonous snakes, such as that vipers will drink wine and shatter glass and that their heads, when boiled, supply an antidote to their venom.

See also: Experiment 17, The "handedness" of molecules, 1848 (see page 76); Experiment 41, The microscopic observations of microbes, 1670s (page 170).

Organizers that control development (1921-1924)

Q How do embryos acquire their shape and form?

Embryos undergo a remarkable transition as they grow. Beginning as a featureless ball of cells, they spontaneously acquire a structure that becomes the head, limbs, organs, and other features. Where does this patterning come from?

Having proposed in 1900 that the "factors" that determine inheritance reside in the chromosomes of the sex cells, the German biologist Theodor Boveri turned to the question of how the body plan emerges in an embryo. He worked with sea urchins: a convenient organism to study, being simple enough for conducting experiments yet complex enough that such studies might be expected to have some relevance to humans and other mammals. Boveri argued that the spherical symmetry of the embryo was broken at the outset by some substance in the egg that came from the mother, and which was concentrated in one region.

This basic idea that embryo growth—embryogenesis—is governed by biological molecules that diffuse through the embryo became popular in the early twentieth century. It was pursued by Boveri's student Hans Spemann, who began to identify the different tissues in the early embryo that would develop into the specialized forms of the adult organism, such as nerves, skin, muscle, and organs. Working at the University

HANS SPEMANN | 1869–1941

Born in Stuttgart, Hans Spemann was one of the generation of naturalists for whom the study of life still held the allure it acquired from the German Romantic tradition of *Naturphilosophie*, championed by the likes of Friedrich Schelling and Johann Wolfgang von Goethe. After studying medicine at Heidelberg, he worked with the cytologist Theodor Boveri and the physicist Wilhelm Röntgen, discoverer of X-rays, at Würzburg. He became an expert at manipulating embryos to alter their development: experiments that are now seen as precursors to cloning technology.

Hans Spemann working in his laboratory, *Embryo Project Encyclopedia* (1931), Marine Biological Laboratory Archives, The University of Chicago.

A photograph of German biologist Hilde Mangold, taken in the early 1920s, private collection.

of Freiburg with amphibian embryos, Spemann found that transplanting a piece of an embryonic tissue called the mesoderm to a different part of the organism could trigger the out-of-place growth of a spinal column in the new location.

In the late 1910s, Hilde Mangold was studying biology and zoology at the University of Frankfurt, and decided to focus on embryology after hearing Spemann speak on the topic. She began her doctoral work in 1920 in Spemann's lab in Freiburg. Spemann did not seem to show much faith in the abilities of women in science, and he assigned Mangold a rather boring dissertation topic. But she proved herself to be a very able student and convinced her supervisor to let her study a more challenging question: how embryonic cells are assigned to specific tissue types.

Spemann and Mangold investigated how this process happens during the crucial period of embryo growth called gastrulation, when the blob of cells first starts to acquire the axis that will become the backbone and spinal column: the key symmetry axis of the body. Spemann's earlier transplantation work suggested that some cells were taking instructions about identity from others that acted as "organizers." The organizer cells responsible for defining the axis that appeared during gastrulation seemed to reside in the so-called dorsal area, along the nascent backbone region at the rear of the body. If that were so, they reasoned, then transplanting dorsal cells to another part of the embryo might trigger the development of another body axis.

Conducting this process in amphibian embryos was a very delicate business. The outer membrane of the egg had to be removed without damaging the embryo, and the cells were manipulated using very fine glass needles. Spemann had devised another

An original slide from Hilde Mangold's experiments on amphibian embryos in the search for organizer cells, 1920s, Berlin Museum of Natural History, Germany.

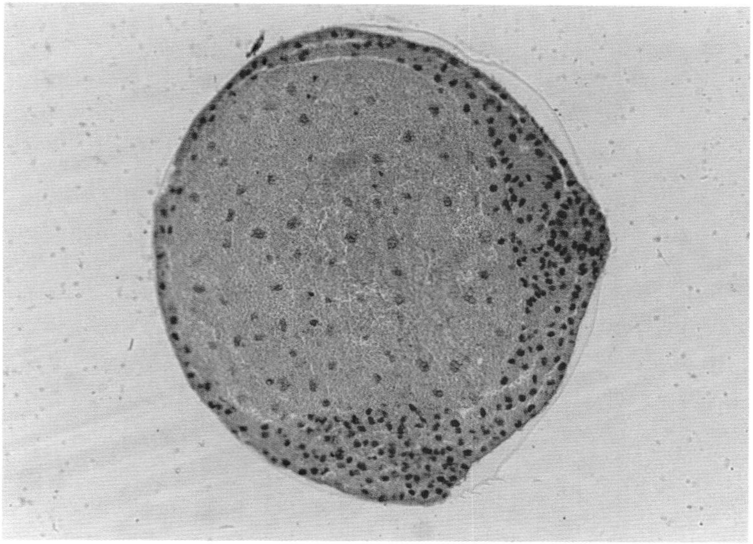

instrument for moving cells around too: a kind of lasso made from the very fine hair of a young child formed into a loop at the tip of a glass tube.

Mangold had to master this fine art in order to remove dorsal cells from salamander embryos and press them onto the other ("ventral") side of the embryo, where they would hopefully stick and grow as a graft. She dyed the transplanted cells so they could be easily identified, returned the embryos to a pond and watched what happened.

Without the protective sheath of the egg membrane, most embryos died from infection. But a few survived—and after about two days of growth, Mangold saw that indeed they developed a second body axis, which could even grow into a second nascent spinal column and brain. Over two years, Mangold saw this happen in a few cases after performing the transplant in around 250 amphibian embryos. She had proved that embryonic development is indeed sometimes controlled by clumps of organizer cells.

Although Mangold did all the experimental work, her thesis paper published in 1924 carried Spemann's name as the first author. Sadly, she was never to enjoy any recognition for her work, dying in an accident just after finishing her doctorate. When Spemann was awarded the Nobel prize in medicine or physiology in 1935—surely in part for the discovery of embryonic organizers—he made only fleeting reference to Mangold in his address.

HILDE MANGOLD | 1898–1924

Hilde Mangold, née Proescholdt, was the daughter of wealthy industrialists in Gotha Thuringia in eastern central Germany, who ran a soap factory. After studying at Frankfurt, she moved to Freiburg to work with Spemann. She had only just completed her doctorate in 1924 when she moved to Berlin with her husband Otto Mangold (the chief assistant in Spemann's lab) and their young son. There she was so severely burned by the explosion of a kitchen gasoline heater that she died of her injuries.

See also: Experiment 44, The role of sperm in fertilization, 1777–1784 (page 180); Experiment 53, The cloning of Dolly the Sheep, 1996 (page 211).

The random nature of genetic mutations (1943)

Q Do genetic mutations occur randomly, as Darwin's evolutionary theory supposes?

By the early 1940s, the marriage of Darwin's theory of evolution by natural selection with the modern understanding of genetic inheritance seemed complete and secure. The idea was now that genes were subject to random mutations in their molecular structure, and the corresponding variations in phenotypes (characteristics and traits) that they produced in organisms were subjected to the winnowing influence of selection. This picture was summarized in a 1942 book by Julian Huxley, grandson of Darwin's staunch advocate Thomas Henry Huxley, which he called *The Modern Synthesis*.

Still, there were weak points in this picture. For one thing, although most researchers assumed genetic mutations do indeed happen at random, there was no direct proof that this was so. There was no reason anyone knew of why a beneficial mutation might not happen *in response to* some stress, which might then be inherited by the offspring of those organisms that survived it. This would be a type of Lamarckian evolution: the inheritance of a characteristic acquired in the organism's lifetime, as proposed in the early nineteenth century by French naturalist Jean-Baptiste Lamarck. Some researchers suspected that this was highly likely for bacteria, which were still something of a mystery. It was not even generally accepted in the mid-twentieth century that bacteria have genes like ours at all; Huxley excluded them from his *Modern Synthesis*.

All the same, bacteria are an excellent subject for investigating natural selection. Their rapid multiplication by simple cell division from a single cell means the consequences of selective pressure can be followed over many generations within the course of just a few days. In 1943, the German–American biologist Max Delbrück and the Italian Salvador Luria set out to establish in what manner mutations really do occur in bacteria.

Delbrück trained in Germany as a physicist, where his research on X-rays and gamma rays led to an interest in their ability to induce genetic mutations. His studies of X-ray-induced mutation, with geneticist Nikolai Vladimirovich Timoféef-Ressovsky and radiation physicist Karl Zimmer in the mid-1930s, allowed them to deduce that genes were of molecular size, and they proposed that mutations involve the alteration of chemical bonds

MAX DELBRÜCK | 1906–1981

Having studied theoretical physics at the University of Göttingen, Max Delbrück began working on high-energy radiation like gamma rays while assisting the physicist Lise Meitner in Berlin. His family was active in the resistance to Nazism, and two of his brothers-in-law were executed for participating in the plot to assassinate Adolf Hitler. Following his move to biology and to the United States, Delbrück became an American citizen in 1945. For his studies with Salvador Luria on the genetics and replication of viruses, he was awarded the 1969 Nobel prize in physiology or medicine.

See also: Experiment 49, The proof that DNA is the genetic material, 1951–1952 (page 197); Experiment 50, The copying of genes, 1958 (page 200).

Max Delbrück (standing) and Salvador Luria examining a petrie dish at Cold Spring Harbor Laboratory, ca. 1941, "Profiles in Science," US National Library of Medicine, courtesy of Cold Spring Harbor Laboratory Archives, Long Island.

within such an entity. These experiments played a central role in the influential 1944 book *What Is Life?* by another physicist who turned his attention to biology, the Austrian Erwin Schrödinger.

Delbrück left Nazi Germany in 1937 to work at the Rockefeller Foundation in New York, before moving to Vanderbilt University in Tennessee in 1939. He also spent some time at the California Institute of Technology where, as well as researching the genetics of fruit flies, he studied bacteria and the viruses that can infect them, called bacteriophages (or simply phages). He became convinced that viruses represented the simplest form of organism and, as such, supplied an ideal model system for studying the basic molecular processes of life.

At Vanderbilt, Delbrück began collaborating with Luria, who worked at Indiana University, on the resistance of bacteria to phage infection. They realized that the way such resistance was acquired could provide a way to test whether the Darwinian assumption of random mutation was correct.

The 1943 experiment by Delbrück and Luria is widely regarded as one of the most elegant in all of biology (even if the paper in which they described their results is sadly not). They supposed that bacteria that survive a typically lethal viral infection do so because some genetic mutation confers resistance by an unknown mechanism (possibly involving a molecular change at the bacterial cell surface that repelled the virus). The question was: did the virus *induce* that mutation, as the co-discoverer of phages, French microbiologist Félix d'Hérelle, had suggested in 1926? Or was the resistance mutation present already in the bacterial colony before infection, simply because such changes in genes happen randomly all the time?

Delbrück and Luria studied infection of the gut bacterium *Escherichia coli* by a bacteriophage called T1. They realized that the two possible sources of the resistance mutation—randomness, as per Darwin, or acquired by "experience," as per Lamarck—would produce different distributions of resistant strains among many infected colonies. If the "resistance gene" had already appeared by chance as a colony grew, it would be conveyed to all successive generations, and that colony would already be dominated by resistant cells. In that case, one would expect that, when many separate colonies were infected with phage, most would be wiped out, but occasionally there would be a colony that had "hit the jackpot," having the resistance gene already. If, on the other hand, it turned out that there was a small and uniformly distributed chance of a bacterial cell finding the wherewithal to survive the virus—and would then pass on this robustness, once brought to light, to its own offspring—then one would expect to find a few little pockets of resistance scattered randomly among many colonies. So in the first case, resistance would be concentrated in just a few colonies, whereas in the latter it would be spread thinly between many.

These possibilities could be distinguished by looking at the statistics of resistance among many separate colonies. That's what Delbrück and Luria measured. They found that resistance followed the "jackpot" distribution, which denoted the presence of pre-existing resistance due to random mutations, rather than the Bell-curve-like distribution expected for acquired mutation. "We consider the results," the pair wrote, "as proof that in our case the resistance to virus is due to a heritable change of the bacterial cell which occurs independently of the action of the virus"—just as Darwinian natural selection supposed. What's more, the results enabled Delbrück and Luria to make one of the first estimates of the rate at which mutations spontaneously occur. This is a crucial parameter in evolution, since it determines how fast an organism can evolve. If the mutation rate is too high, inherited genetic information can't be passed on reliably; if it is too low, it is hard for an organism to adapt to new circumstances. Mutations rates vary between species (and even between different parts of the genome), with viruses having some of the highest, allowing them to evolve quickly to evade a host organism's immune defences.

The results of Delbrück and Luria not only placed the assumptions of Darwinian theory on a firm footing but also stimulated interest in the mechanisms of inheritance and genetics that spurred the growth of molecular biology in the 1950s and '60s. All the same, debate remains about the processes by which mutations happen, and there is some indication that organisms (including *E. coli*) can alter their mutation rates (by changes to the molecular processes of DNA repair and replication) in response to stress, tuning their ability to evolve their way out of trouble.

SALVADOR LURIA | (1912–1991)

Born and educated in Turin, Salvador Luria studied radiology in Rome in 1937, where he encountered Max Delbrück's work on induced mutations. With Jewish heritage, he fell foul of Mussolini's antisemitic laws restricting academic appointments. He left Italy for Paris, but fled to the United States in 1940 as France fell to the German invasion. At Indiana University, his first graduate student was James Watson, who later solved the structure of DNA with Francis Crick.

The proof that DNA is the genetic material (1951-1952)

Q What is the molecular carrier of genes?

In the early twentieth century, it was widely thought that the molecule that encodes genes is a protein. After all, these molecules seemed to be at the heart of biology: as enzymes they enable the chemical reactions involved in metabolism. What's more, when biochemists analyzed the chromosomes—the cell structures that seemed to contain the genes and which were copied when cells divide—they found plenty of protein there. The other component of chromosomes, a kind of biological polymer called deoxyribonucleic acid (DNA), was thought to be some kind of packaging.

ALFRED HERSHEY | 1908–1997

As a bacteriologist at Washington University in St. Louis, Missouri, Alfred Hershey gained his expertise in working with bacteriophage when he collaborated with Salvador Luria and Max Delbrück in the 1940s: the three men were later dubbed the Trinity of the "Phage Church." He joined the Carnegie Institution of Washington's department of genetics in Cold Spring Harbor (later the Cold Spring Harbor Laboratory) in 1950, and became its director in 1962.

See also: Experiment 48, The random nature of genetic mutations, 1943 (page 194); Experiment 50, The copying of genes, 1958 (page 200).

That assumption was challenged by the experiments of microbiologist Oswald Avery, working at the Rockefeller Institute Hospital in New York. Avery was an expert on pneumococci, the bacteria that cause pneumonia. In the 1920s and '30s, he began to study how some of these bacteria could be transformed from a relatively mild form (denoted R, because their cells were rough) to a more virulent form (S, with smooth cells). He was able to extract from the S strain a substance he called the "transforming principle," which seemed able to turn R pneumococci virulent. What was it made from?

Working with Canadian physician Colin MacLeod, Avery used a centrifuge to separate the extracts of cells according to their density. By 1942 the pair had accumulated good evidence that the transforming principle—which seemed able to "reprogram" the bacteria so that they acquired new properties—consisted solely of DNA. In 1943 Avery wrote that if he was right, nucleic acids (DNA and its chemical relative RNA) are "functionally active substances in determining the biochemical activities and specific characteristics of cells … Sounds like a virus—may be a gene." He presented the experimental evidence for these conclusions in 1944.

Strangely (in retrospect), many of those working on the mechanisms of inheritance were relatively unconcerned about that possibility. As Max Delbrück put it, if genes weren't encoded in proteins, then "genetic specificity was carried by some goddamn other molecule." Who cared what it was?

That view was initially shared by American geneticist Alfred (Al) Hershey, an expert on phages: the viruses that infect bacteria. But such indifference to the chemistry of replication and

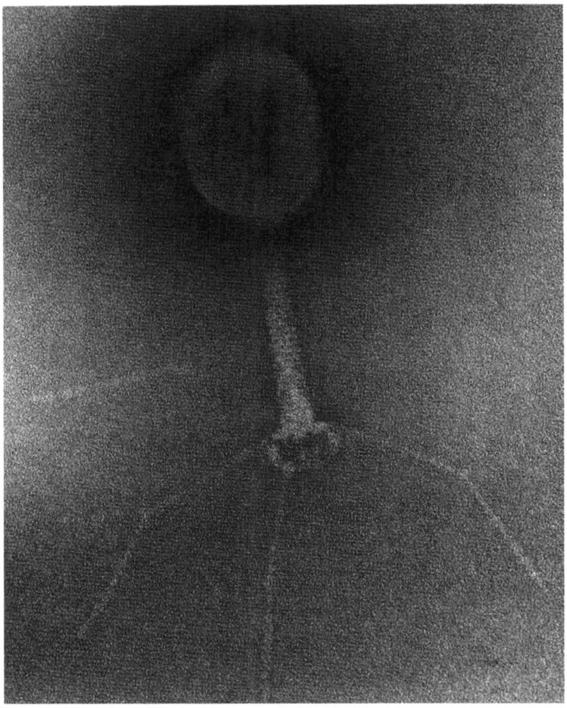

Digitally colored transmission electron micrograph (TEM) of a bacteriophage (a bacteria-eating virus), Wellcome Collection, London.

inheritance couldn't persist if you really wanted to know what was going on in that process. In particular, Hershey wished to understand how viruses work: how they hijack the cells they infect so as to replicate themselves. By the late 1940s, studies using the electron microscope had shown that viruses first attach to the outside of cells. But what happens then? Might it be that the virus itself doesn't enter the cell, but rather, that it simply injects something like Avery's "transforming principle"—which might be genetic material—through the cell membrane?

In the early 1950s, Hershey began a series of experiments, in collaboration with his technician Martha Chase, to answer this question. Looking at a phage called T2 that infects *E.coli*, they established that these bacteriophages do indeed inject DNA into the cells to which they attach—leaving behind the protein shell, called "ghost phage" because it was itself inactive. They key question was whether the viral DNA alone is responsible for the effect on the host cell.

Hershey and Chase faced the same problem as Avery and MacLeod: proteins and DNA tend to be intimately associated, so it is hard to separate their potential influence. To do that, Hershey and Chase devised an ingenious ploy. They used radioactive atoms to label the protein and DNA components of the phage, so that where the radioactivity ends up could reveal where the respective molecules were. By letting the phage replicate in bacteria grown in a culture medium that contains a radioactive form of sulfur (denoted ^{35}S), they could make a form of the virus in which the proteins incorporate the telltale sulfur atoms (present in two of the amino acids that make up proteins). There is no sulfur in DNA, so that remained non-radioactive. But by using a radioactive type of phosphorus (^{32}P) instead, they could label the viral DNA with radioactivity, since only that contains phosphorus.

When *E. coli* were infected with the ^{35}S-phage and the ghost phage was then separated from the cells using a centrifuge, there was almost no radioactivity detected in the infected cells: it was

MARTHA CHASE | 1927–2003

When Martha Chase, born in Ohio, came to work at the Cold Spring Harbor Laboratory in 1950, she did not yet hold a doctoral degree; she gained that in 1959 at the University of Southern California. She left Cold Spring Harbor soon after her work with Alfred Hershey in 1953, but remained a member of the informal "Phage Group," centered around Max Delbrück, to which the beginning of molecular biology is often credited.

found that little of the protein was transferred to them. (In truth this separation wasn't clearcut—about 20 percent of the protein remained.) But when ^{32}P-phage was used, not only did the cells become markedly radioactive but some of this radioactivity was passed on to the phage replicated within the cells.

Hershey and Chase concluded that the protein component of phage is just a protective coat that enables the virus to attach to cells and inject material: it plays no part in the growth of phage inside the cells. The phage DNA, meanwhile, is injected inside and has some function in that replication process.

The results of these experiments are often presented as confirmation that the genetic material—the stuff viruses pass into cells—is DNA. Hershey and Chase were much more circumspect, however. Not only did they not rule out the possibility that some other material besides DNA (but which contains no sulfur) might also be transferred to infected cells, but they also wrote that "the chemical identification of the genetic part" of what viruses transfer remains an open question. Even by 1953 Hershey himself insisted there was not enough evidence to conclude that the genes are made of DNA alone. He said that biologists remained equally divided on that issue, and he himself continued to suspect that protein might be involved.

The experiments of Hershey and Chase, for which Hershey shared the 1969 Nobel prize with Max Delbrück and Salvador Luria (Chase was excluded, her contributions not even acknowledged in Hershey's acceptance speech), far from clinching the debate, therefore illustrate how advances in scientific understanding usually happen: little by little, with great caution and some backsliding, and usually retaining a residue of old ideas that the experiments allegedly discredit. It generally takes more than one new experimental finding to shift deeply entrenched preconceptions. And perhaps that's how it should be.

Alfred Hershey (right) and Martha Chase in 1953, courtesy of Cold Spring Harbor Laboratory Archives, Long Island.

The copying of genes (1958)

Q How does double-stranded DNA replicate?

When James Watson and Francis Crick described the double-helical structure of the DNA molecule in 1953, they ended their epoch-making paper in the journal *Nature* with a famous observation: "It has not escaped our notice that the specific pairing [of strands] we have postulated immediately suggests a possible copying mechanism for the genetic material." What Watson and Crick—whose work drew heavily on the insights provided by Rosalind Franklin's crystallographic studies of DNA—had in mind was that, to make a copy of itself, the two strands of the double helix could separate and each act as a template for constructing a new partner strand. The twin strands in DNA are held together by weak yet selective chemical bonds between molecular units called nucleotide bases that dangle from the backbone of the strand. There are four of these bases, denoted C, G, A, and T, and the rule is that C sticks to G and A to T. Thus, a single strand can template the assembly of a new strand with the complementary sequence of bases.

This is indeed what happens when DNA is replicated before a cell divides, so that each of the daughter cells has a full copy of all the gene-carrying chromosomes. Such a process is said to be semiconservative, meaning that each of the two new double helices contains one strand from the original "parent" DNA and one made afresh by using it as a template.

Yet despite Watson and Crick's casual confidence in this principle of DNA replication, not everyone agreed that the semiconservative model was the only one possible. Max Delbrück at the California Institute of Technology (Caltech) argued that it was no easy matter for the double helix to unwind and the strands unzip, as Watson and Crick supposed. In 1954 he suggested instead that DNA was copied by small pieces breaking off and acting as templates for duplication, before the fragments were reassembled as a kind of patchwork of new and old: an idea called the dispersive model. Later he and the biochemist Gunther Stent suggested yet another possibility: a completely new double helix might be formed on the original one, so that the parent helix was wholly preserved. This was called a conservative model.

Who was right? In 1954, graduate student Matthew Meselson at Caltech talked with Delbrück

MATTHEW MESELSON | B. 1930

One of the many scientists to cross over from physical to biological sciences during the advent of molecular biology, Matthew Meselson was a graduate student of the great chemist Linus Pauling at the California Institute of Technology during the mid-1950s. It was there that he studied X-ray crystallography, which Pauling and others were starting to apply to complex biomolecules such as proteins. From the 1960s Meselson worked extensively on the control and prohibition of chemical and biological weapons.

Matthew Meselson at the controls of the UV optics and photography system used for Meselson and Stahl's DNA replication experiment, 1958.

about the problem of DNA replication, which was a topic of his PhD. Later that year, Meselson met another postgraduate working on the same problem: Franklin Stahl, who was based at the University of Rochester in New York. They agreed to work together on the matter and began collaborating at Caltech in 1956.

The problem faced by Meselson and Stahl is that when DNA is replicated, by definition two identical copies are made from the original one. If the parent and daughter strands are chemically indistinguishable, how then can we tell which is which? The two researchers found a beautifully simple solution: make them different. Their first idea was to start with "parent" DNA that was chemically modified. If the conservative model was right, a round of replication would end up with one modified double helix and one normal double helix, the latter being the daughter product. If replication was dispersive, you'd get total patchworks of modified and unmodified fragments in the daughters. If it was semiconservative, each daughter double helix would be composed of one modified strand and one unmodified.

How, though, to tell them apart? Meselson and Stahl figured that if the parent DNA had some extra atoms attached—they initially tried adding bromine atoms—then it would be heavier

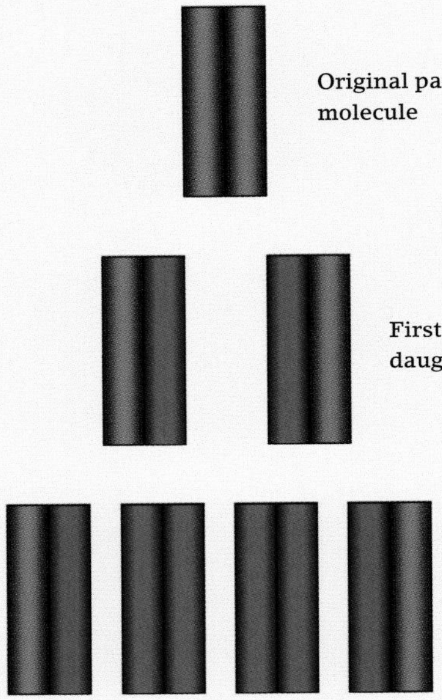

An interpretation of what the density labeling data confirms in terms of a model for DNA replication. From a journal article by Matthew Meselson and Frank W. Stahl in the *Proceedings of the National Academy of Sciences* (PNAS), USA, July 15, 1958, 44(7), Fig. 5.

FRANKLIN STAHL | B. 1929

With a degree in biology from what was then Harvard College, Franklin Stahl joined the efforts to understand the molecular basis of life via the study of bacteriophage viruses and their effects on bacteria. He studied these systems at the Cold Spring Harbor Laboratory and the California Institute of Technology, and after his seminal work with Meselson he continued working on phages at the University of Oregon until his retirement in 2001.

and denser than normal DNA. They could then use a technique devised by Meselson—a form of centrifugation, in which substances are rapidly spun so that the densest are pushed to the bottom—to separate the different types of DNA made by replication and see what they found. They began by using DNA from a type of virus, but soon switched to the DNA of the bacterium *Escherichia coli*, which didn't get replicated as fast and so allowed them to study the products one replication cycle at a time.

However, it was difficult to get enough bromine incorporated into the parent DNA to make it significantly denser than normal DNA. In late 1957, they adopted a different approach: making the parent DNA heavier by using different isotopes. Every chemical element has different isotopic forms, in which the atoms have different masses because their nuclei contain different numbers of neutrons. Meselson and Stahl grew their *E. coli* in solutions in which the nutrients contained two different isotopes of nitrogen (denoted ^{14}N, the isotope that almost totally dominates in air and in natural substances, and ^{15}N), which would be incorporated into the DNA they made. If the initial parent DNA contained only ^{15}N, but subsequent replication cycles were conducted in a medium containing the lighter ^{14}N, the daughter strands would be lighter. The mass difference is only small for individual nitrogen atoms, but is cumulatively significant as DNA contains many of them: enough to separate ^{14}N-DNA and ^{15}N-DNA by centrifugation.

Extracting and analyzing the DNA before any replication showed that indeed it contained only ^{15}N. As the first replication cycle proceeded, fed by only ^{14}N-containing ingredients, this fraction disappeared and DNA of a lower density appeared. If replication was conservative, there would be two products of different density: one with ^{15}N in both strands, the other with just ^{14}N. If it was dispersive, meanwhile, there would be a whole range of different densities. But the semiconservative model would produce a single type of daughter DNA, with ^{15}N in one strand and ^{14}N in the other. That's what the two researchers found. Further rounds of replication also produced the distribution of nitrogen isotopes predicted by the semiconservative model—in particular, the second cycle produced some DNA with double strands of ^{14}N alone, which was more or less impossible from the dispersive model.

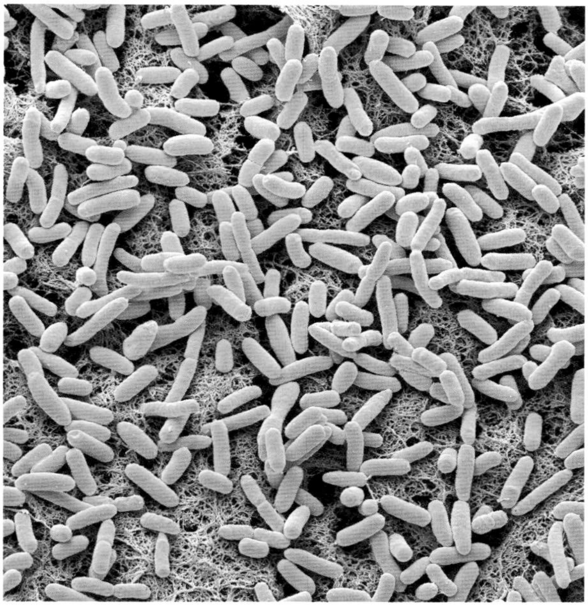

The human gut bacterium *Escherichia coli*, used by Meselson and Stahl in their experimental study of how DNA is replicated.

To his immense credit, once Delbrück heard about these results, he accepted at once that the semiconservative model was the right one, despite having challenged it so strongly previously. It was, after all, hard to dispute: the experiment was so elegant, and the findings so clear, that Meselson and Stahl's study has been called "the most beautiful experiment in biology." Not only did it resolve the dispute but, in doing so, it elucidated one of the most fundamental processes of living things.

The proof that the genetic code is a triplet code (1961)

Q How many DNA bases code for a single amino acid in a protein?

Once it had become clear that the genetic code represented each amino acid in a protein as a group of nucleotide bases in the DNA (and messenger RNA) of the corresponding gene, then the race was on to crack this code. Most researchers assumed that these groups, which are called codons, are triplets, because that is the smallest number that can give enough permutations of the four bases (denoted C, G, A, and U in RNA) to encode all twenty of the amino acids found in proteins.

FRANCIS CRICK | 1916–2004

Like many of the scientists who pioneered molecular biology in the twentieth century, Francis Crick trained as a physicist (at University College London), working on mine technologies for the British Admiralty during the war. He subsequently took an interest in biological questions at the University of Cambridge, where he and James Watson worked out the chemical structure of DNA in 1953, while Crick was still completing his PhD on the crystallography of biomolecules. Toward the end of his career, Crick turned his attention to neuroscience and consciousness, promulgating a thoroughly materialistic view of the mind. One of the most perceptive and creative scientists of his time, Crick did not escape some of the prejudices of his upbringing, remaining a committed eugenicist all his life.

But there was no proof of that, and no reason nature *had* to do things this way. It was perfectly possible that codons contained four or more bases. There was also the question of how they were arranged. Might codons overlap, for example? That was rendered unlikely by experiments at the start of the 1960s showing that when viruses were treated with chemical agents known to damage RNA bases, such alterations to a base sequence tended only to change a single amino acid in its protein product: each base was part of only one codon. Other questions remained. Did the codons just follow one another in sequence, or might they be separated by, say, a single base that acts like a comma, or perhaps by entire "nonsense" sequences that are ignored as the genetic information is read out?

In 1961 Francis Crick, working with South African geneticist Sydney Brenner at Cambridge University, figured out an experimental way to explore the true nature of codons. They, too, used chemical agents to damage the genetic material (in this case, RNA) of a virus—the T4 bacteriophage, which had become the "workhorse" of genetic studies—and looked at the consequences for its ability to infect *Escherichia coli* bacteria. They focused on a region of the viral genome that contained just two genes.

Their RNA-damaging ingredient was a form of a synthetic yellow dye called acridine, which can stick to a base and knock it out of action—in effect deleting a base. However, the acridine molecule could also *add* a base pair into the RNA sequence. There was no telling which it might do—but either way, adding or deleting a base would throw the molecular machinery translating codons to proteins out of step. For example, suppose the codons are indeed sequential triplets, and we start

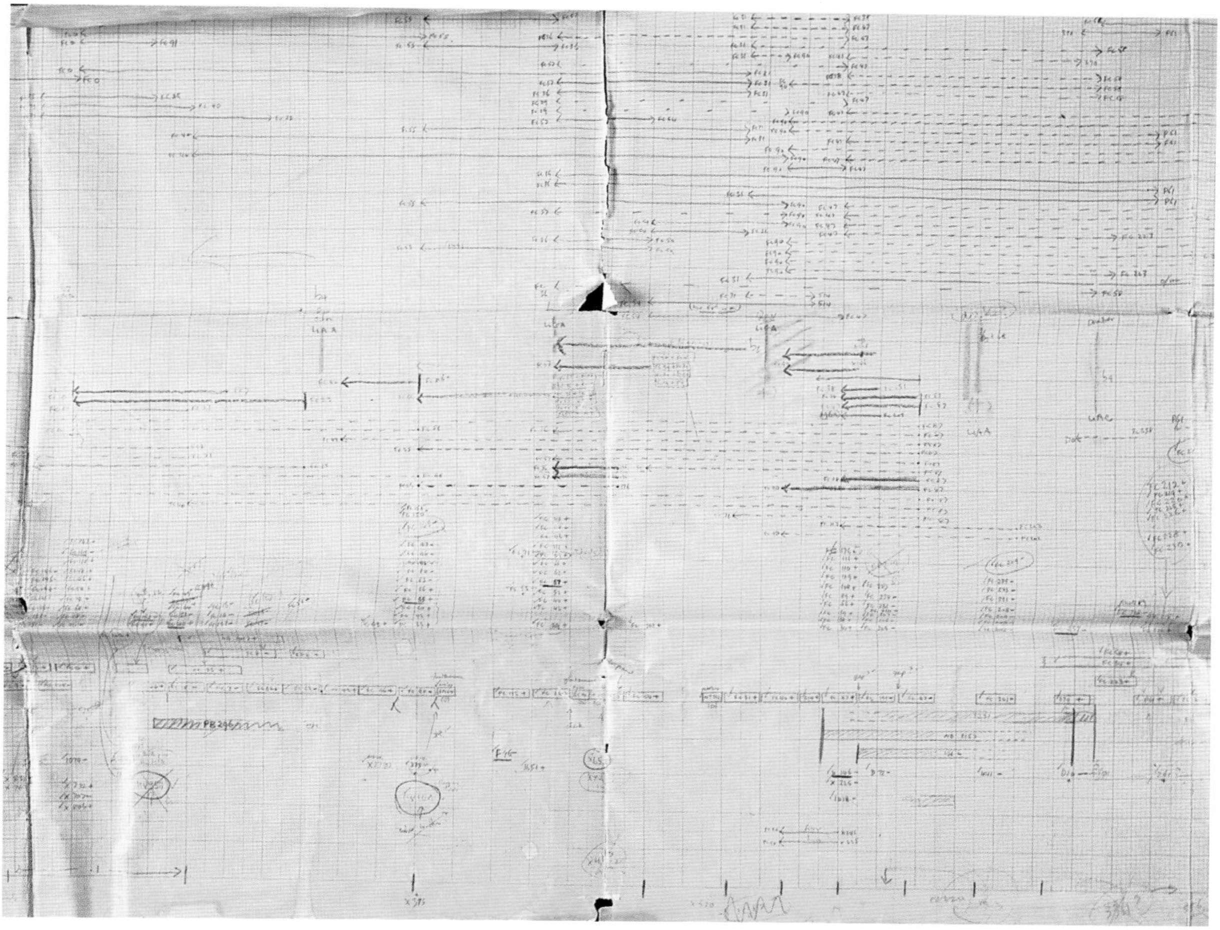

Section from a draft of a genetic map used by Francis Crick, Sydney Brenner, and their coworkers, the Crick papers, the Triplet Code, 1961, Wellcome Collection, London.

with a sequence ATG-CAT-CCC-TGA. If the first C is deleted, then from this point on the sequence is read as ATC-CCT-GA ... In other words, the "reading frame" is shifted—and the corresponding protein produced will almost certainly be "nonsensical" and non-functional. If it is a protein crucial to the virus's survival, this frame shift will destroy the virus's ability to infect bacteria. The same is true if a base is added to the sequence.

Crick and colleagues found that indeed these experiments in gene damage produced inactive phages. Suppose, however, that a second point of damage were to be made a little farther along the RNA strand—and that it has the opposite effect to the first, for example adding a base where the first attack deleted one. Then the proper reading frame would be restored from that point on. So the corresponding protein might have a short faulty segment, but would be mostly faithfully synthesized—and with luck, its function might survive. The researchers did indeed find that they could restore function to "single-mutation" strains this way.

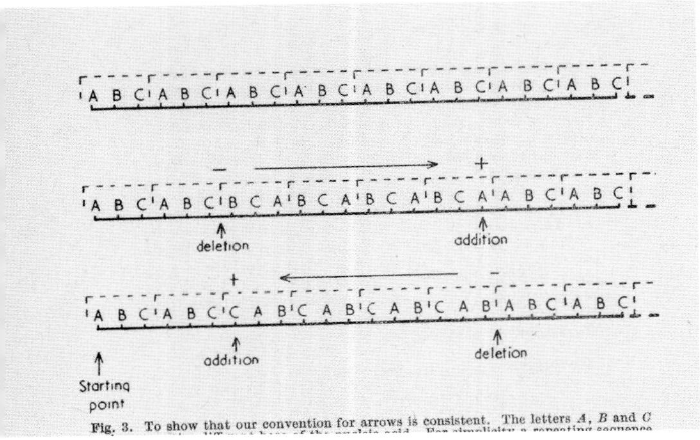

The triplet nature of the genetic code. From Crick, Barnett, Brenner, and Watts-Tobin's "General nature of the genetic code for proteins," *Nature*, December 30, 1961, Vol. 192, no. 4809, Fig. 3. Wellcome Collection, London.

This would all be true regardless of whether the codons were groups of 3, 4, 5, or more bases. But the team then looked at *triple* mutants, in which three bases had been altered. For cases where all three damage points involve either addition or deletion of a base, the reading frame would be restored if, and only if, codons are triplets. This is just what Crick, Brenner, and their colleagues found.

The experiments were challenging. It was no mean feat to combine viral mutations in such a way that they could ensure that a strain had precisely three additions or deletions of bases. Those studies, requiring the researchers to look at the effects of about eighty different mutant forms of the virus in their colonies of *E. coli*, were done by Crick and microbiologist Leslie Barnett in the fall of 1961, while Brenner was away on a visit to Paris. By late September they had their answer. After checking the bacterial growth late one night, Crick said to Barnett: "We're the only two to know it's a triplet code!"

The researchers published their findings in *Nature* on December 30. They also reasoned that the code must be "redundant": at least some of the twenty amino acids would be coded by more than one of the sixty-four possible triplet codons. As a result, different organisms might favor different codons for the same amino acid.

These experiments operated, as ground-breaking experiments often must, right at the edge of what the techniques can handle. Frankly, such studies depend on a considerable dash of good fortune: the phenomena are often so poorly understood, and the instrumental methods so rudimentary, that only in rare cases lacking hidden complications will they work at all. Crick and colleagues had no idea quite how their reagent was damaging the target gene (that's still not entirely clear today). They didn't know what that gene did, nor even if it actually produced a protein at all, let alone one crucial to the viability of the virus. They had to hope for the best. A part of the art of experimentation often comes from good intuition

LESLIE BARNETT | 1920–2002

Trained as a microbiologist, Leslie Barnett began her career working on prosaic issues such as bacteriological testing of milk for the dairy industry. She became a technician at the Medical Research Council's Cambridge laboratory, initially working with the lab's computers; only when geneticist Sydney Brenner arrived at Cambridge were her biological skills called on. She trained Crick to do experiments with bacteriophages and later worked with Brenner at Addenbrooke's Hospital before becoming a Fellow and tutor at Clare Hall, Cambridge.

See also: Experiment 50, The copying of genes, 1958 (page 200); Experiment 52, Reading the first letter of the genetic code, 1960–1961 (page 208).

about how much ignorance of what is really happening you can get away with.

But if fortune smiled on Crick and colleagues, nonetheless their argument has great elegance: the American biochemist Marshall Nirenberg, who was also working to crack the genetic code, called it "beautiful." In 2004 the original manuscript sent to *Nature* sold at auction for £13,145. With the basic composition of codons now established, Crick said that, so long as the genetic code (the actual correspondence between base triplets and amino acids) remained the same in all organisms, it "may well be solved within the year." As it transpired, the task was a bit harder than that.

James Watson (left) and Francis Crick in 1953 with their model of a molecule of DNA. They met at the Cavendish Laboratory at Cambridge University.

Reading the first letter of the genetic code (1960–1961)

Q What is the genetic code that relates DNA to proteins?

In November 1958, biochemist Marshall Nirenberg, working at the US National Institutes of Health in Bethesda, Maryland, figured he could see how to deduce the correspondences between codons and amino acids. The key was to find a way of synthesizing a protein from its corresponding messenger RNA outside of living cells, in a test tube. That way, you could specify the RNA sequence that encoded the protein.

MARSHALL NIRENBERG
1927–2010

Marshall Warren Nirenberg grew up in Florida, where his family moved in the hope that the climate would ease his childhood rheumatic fever. Fascinated by the local flora and fauna, Nirenberg studied zoology at the University of Florida at Gainesville, and in 1952 he completed his master's thesis there on caddis flies. But after moving to the University of Michigan for his PhD he switched to biological chemistry, where he was drawn to the problem of the day: the genetic code. Described as painfully modest, Nirenberg was awarded the 1968 Nobel prize in physiology or medicine for cracking the first letter of the genetic code.

See also: Experiment 50, The copying of genes, 1958 (page 200); Experiment 51, The proof that the genetic code is a triplet code, 1961 (page 204).

With such a capability, Nirenberg wrote in his lab book, one "could crack life's code!"

Protein synthesis *in vitro* had been achieved just a year earlier by American biochemist Paul Zamecnik in Boston and his colleagues, who found that the formation of natural proteins from mRNA could proceed among the contents of rat liver cells even after they had split the cells apart: the cell extracts contained all the molecular components needed for the job. (Others working in Zamecnik's team also discovered that the process involved another form of RNA besides mRNA, which became known as transfer RNA.) In 1960 Nirenberg and his German colleague Heinrich Matthaei tried using such cell-free extracts to make protein-like molecules not from the mRNA already present in the cells but from RNA molecules that they supplied themselves, and which they had synthesized using chemical methods.

Today researchers can produce any RNA sequence using biotechnological methods to string its four nucleotide bases in whatever order they choose. In the early 1960s that wasn't possible. But what Nirenberg and Matthaei *could* do was to link just one type of base into a long polymer strand, using a method devised by the Spanish biochemist Severo Ochoa. In work that earned him the 1959 Nobel prize in physiology or medicine, Ochoa discovered that an enzyme called polynucleotide phosphorylase could link together individual nucleotide bases. There was no way to control the order in which the bases were stitched together, but if they were all the same, it made no difference.

Nirenberg and Matthaei figured that, if codons containing just one type of base (such as AAA or UUU) encoded a specific amino acid in a protein, then RNA strands made from just one of those bases (denoted poly(A), poly(U), and so on) would

Marshall Nirenberg (right) and Heinrich Matthaei in the laboratory, in January 1962, National Institutes of Health, Bethesda, Maryland.

act as templates or instructions for the protein-making machinery to churn out long chains composed solely of the corresponding amino acid: each group of three bases would be interpreted as a codon. A chain of just one type of amino acids joined together is not a real protein—it wouldn't have any biological function—but it *is* a molecule called a polypeptide, of which natural proteins are a subset.

By November 1960, the two researchers had mastered the art of cell-free protein synthesis using extracts from the bacterium *Escherichia coli*. Could they get this system to use repetitive strands of RNA made using Ochoa's method as the template for making repetitive polypeptides? If so, it should be possible to figure out which amino acids the same-letter codons encoded, by seeing which ones were used in the polypeptides assembled from the instructions on the RNA.

The researchers could isolate the proteins and polypeptides made by their cell-free system, but lacked any analytical methods to find out directly which amino acids they contained. Matthaei found an ingenious solution. He used amino acids in which some of the carbon atoms were exchanged for the radioactive isotope of carbon, ^{14}C. This is formed naturally in the atmosphere and is incorporated into living organisms. It decays

HEINRICH MATTHAEI | B. 1929

When Marshall Nirenberg began studying the genetic code at the National Institutes of Health in 1960, he was joined by postdoctoral researcher Johannes Heinrich Matthaei, who was based at New York's Cornell University on a visit from the University of Bonn. Although the key radiocarbon-labeling method was partly Matthaei's inspiration, he was excluded from the 1968 Nobel prize for reasons never made clear.

radioactively by emitting a beta particle—which provides a convenient signal of whether it is present in some chemical substance.

Matthaei deduced by a process of elimination which amino acid was being used when he added poly(U) to his system. There are twenty natural amino acids in proteins, and he began with mixtures of ten labeled with ^{14}C and ten with normal carbon. If the products emitted beta particles, they contained one of the ten ^{14}C-labeled amino acids. He could then repeat the process with those ten, dividing them into five labeled and five unlabeled—and so on. In this way, by May 1961 Matthaei had narrowed down the amino acid used by poly(U) to just two: either phenylalanine or tyrosine. The final test involved feeding the cell-free protein synthesis system with just one of each of these labeled with ^{14}C. This way, he was able to show that poly(U) made long chains of phenylalanine—implying that the poly-U codon coded for this amino acid. When he and Nirenberg reported their findings in October, they were careful to admit that they didn't yet know for sure that this codon was a triplet (as widely suspected) or a larger group.

Nirenberg's success took everyone by surprise, because he was not one of the "big shots," like Max Delbrück or Francis Crick, who were thinking about the problem of the genetic code. According to Crick, his announcement of the result at an international meeting in Moscow that August left the audience "startled" and "electrified." In retrospect the experiment using synthetic polynucleotides seemed so obvious. In fact, a few others (including researchers in Ochoa's lab) had been trying it without success, and Matthew Meselson attested that the first instinct of some of those at Nirenberg's Moscow talk was to get back to their labs at once and modify his methods to crack the rest of the code. By 1967 they had done so for all sixty-four triplet codons, effectively closing the first chapter of modern genetics. As Crick put it in 1962, "If the DNA structure was the end of the beginning, the discovery of Nirenberg and Matthaei is the beginning of the end."

Marshall Nirenberg holding DNA models at the National Institutes of Health in the 1960s, National Institutes of Health, Bethesda, Maryland.

The cloning of Dolly the Sheep (1996)

Q Can a mammal be cloned from the genetic material in a mature adult cell?

The possibility of cloning has been long imagined. The word derives from the Greek for "twig," alluding to the propagation of plants from cuttings. In 1959 the French scientist Jean Rostand imagined this idea extended to humans: if we took a piece of tissue from a freshly dead body, he said, we might imagine "the perfected science of the future as capable of remaking from such a culture the complete person ... This would, in short, be *human propagation from cuttings*."

IAN WILMUT | B. 1944

Having studied agriculture at the University of Nottingham in England, Ian Wilmut's interest in cloning was always motivated as much by the practical applications as the fundamental biology. That was reflected, too, in his doctoral work at Cambridge on the cryopreservation of pig and boar semen, and his appointment at The Roslin Institute in Scotland, formerly the Animal Breeding Research Organisation. Although he headed the team that cloned Dolly, Wilmut insisted that much of the credit for the work should go to his colleague, animal biologist Keith Campbell.

See also: Experiment 47, Organizers that control development, 1921–1924 (page 191); Experiment 54, The first "synthetic organism," 2010 (page 214).

It's not that simple, of course: a tissue culture won't spontaneously grow into a complete organism. For one thing, the cells of tissues have already become specialized into a particular type: a heart cell, say, or a neuron.

However, embryo cells in the very early stage of development have not yet narrowed their options. Called embryonic stem cells, they remain pluripotent, meaning that they can develop into all tissue types of the body. The first person to exploit that versatility for cloning was the German biologist Hans Driesch, who back in 1892 shook two- or four-cell sea-urchin embryos until their cells separated, and watched each of the fragments grow into a whole sea urchin.

In the 1900s, Hans Spemann divided two-cell salamander embryos using a lasso made from his infant's hair (see page 191). But it was the experiment conducted by Spemann in 1928 that truly paved the way for the cloning of larger animals. Using fine needles to manipulate the cells, he dragged the compartment called the nucleus, containing the chromosomes, out of a freshly fertilized salamander egg, and replaced it with the nucleus from an embryonic salamander cell. The new set of chromosomes allowed the egg to grow into a complete embryo.

In his Nobel prize speech of 1938 Spemann fantasized about using such "nuclear transfer" to move the chromosomes from a fully matured body (somatic) cell into an egg, so as to clone an adult organism. That experiment was first done, after a fashion, in 1955 by Robert Briggs and Thomas King in Philadelphia, using frog cells. They didn't use an adult cell as the nuclear donor, but rather the cell of an advanced-stage embryo. John Gurdon, working in Cambridge in the 1960s, succeeded in cloning frogs by somatic-cell nuclear transfer (SCNT) from

Dolly with her own first lamb, Bonnie, in April 1998, courtesy of The Roslin Institute, The University of Edinburgh.

cells at a yet more advanced stage of development, although still not from fully adult cells.

In the 1980s sheep, cows, and mice were cloned by SCNT, and in 1995 scientists at The Roslin Institute, a UK government research laboratory in Scotland, made two cloned sheep, which they named Megan and Morag, by SCNT from embryo cells cultured for several weeks. Later that year, they made a quartet of sheep clones using nine-day-old embryo cells as the chromosome donor. Typically, the egg that hosts the transferred nucleus can be stimulated to develop simply by delivering a small electrical jolt.

Given this extensive past history of cloning by SCNT, it might seem odd that the sheep called Dolly, cloned in 1996, made such headlines when she was announced in February the following year by the Roslin team, led by Ian Wilmut and Keith Campbell. But what made Dolly special was that she was the first large mammal cloned from a fully matured adult donor cell: specifically, a cell taken from the mammary tissue of an ewe. (The name Dolly was chosen, with questionable schoolboy humor, in allusion to the singer Dolly Parton.) This meant that for the first time it seemed feasible to imagine cloning an adult human from one of our cells: taking its chromosomes and moving them into an unfertilized egg, and thereby growing a genetic copy of the person.

That prospect has never been realized, for many scientific and ethical reasons. In particular, there is a high failure rate for mammalian cloning: most of the eggs that have received a new nucleus fail to develop into a full embryo when, like an embryo made by in vitro fertilization, they are implanted into the animal womb. In an experiment on cloning of macaque monkeys in China in 2018—the first cloning of a primate—there were only two live births from six pregnancies generated by the implantation of seventy-nine cloned embryos into twenty-one female monkeys. It's not clear how much those odds can be improved, and that alone furnishes a good reason not to try this process in humans.

What's more, it remains unclear whether there are any long-term consequences for the cloned animal. Dolly died relatively young: at the age of six, she was euthanized after tumors were found in her lungs. Although it's not clear that her cancer had anything to do with her origin by cloning, some researchers suspect cloned animals might be prematurely aged: a trace of their "former life" could remain imprinted in the transferred chromosomes. Wilmut and Campbell have said that cloned fetuses are ten times more likely than ordinary ones to die in the womb, and cloned offspring also have an enhanced risk of early death.

The mere prospect of human cloning guaranteed fame for Dolly—or some might say, notoriety. In a profile of Wilmut, *Time* magazine said: "One doesn't expect Dr Frankenstein to show up in a wool sweater, baggy parka, soft British accent and the face of a bank clerk." The German magazine *Der Spiegel* accompanied its report with an image of a battalion of cloned Hitlers—a long-standing, peculiar obsession haunting the notion of human cloning. Wilmut said he "quickly came to dread the pleas from bereaved families, asking if we could clone their lost loved ones."

The fallacy here—that a cloned person would be identical to the "original," or would even offer a kind of immortality—hasn't prevented the appearance of pet-cloning services. For the Roslin researchers the goal was rather about developing a reliable means of making genetically modified livestock, an aim they achieved with Dolly's successor Polly, born in 1997. Cloning of livestock might also offer a way of propagating the best specimens, such as good milk bearers. Human reproductive cloning meanwhile remains scientifically unproven, illegal in most countries, and lacking in any clear motivation.

"Dolly: the cloning of a sheep, 1996." Diagram from *Encyclopaedia Britannica*.

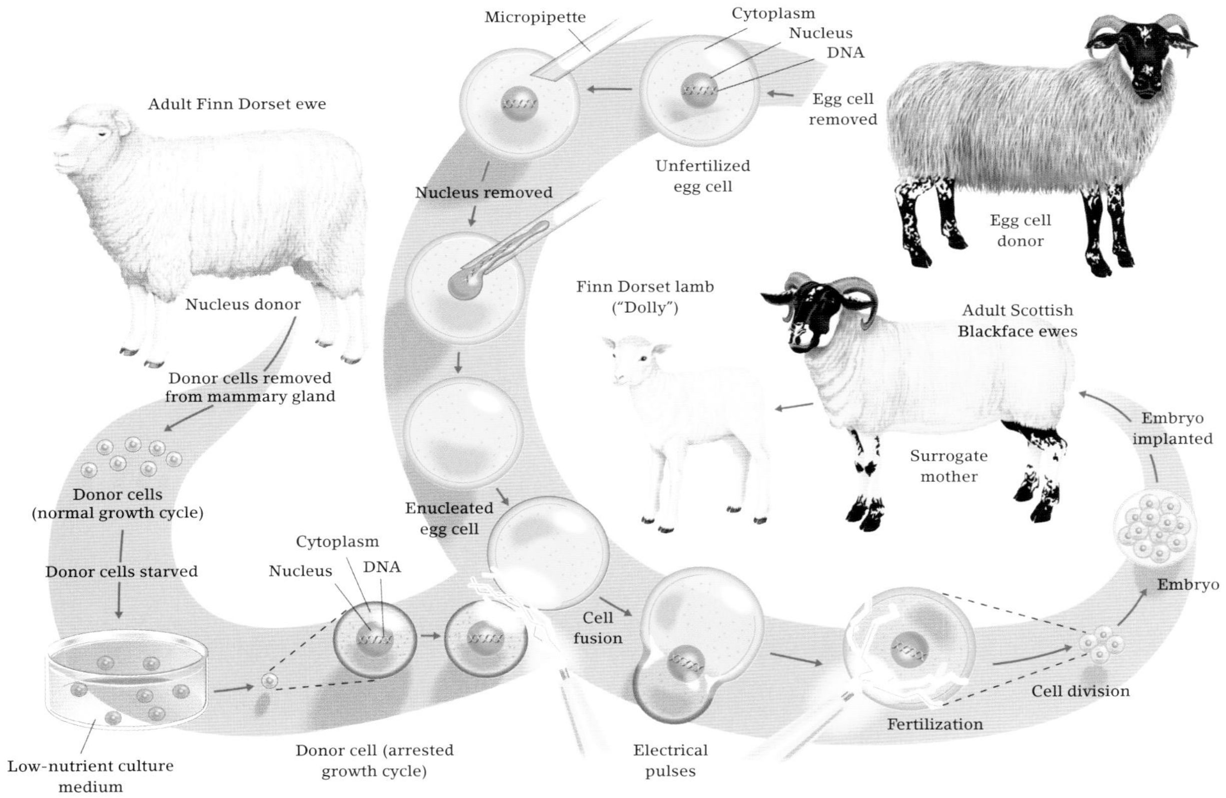

The first "synthetic organism" (2010)

Q Can the genome of an organism be designed and made from scratch?

Biology has long been regarded as a discovery science: the aim is to find out about natural organisms and deduce how they work. But we have for millennia possessed an ability to modify nature too, via selective breeding—a kind of "unnatural selection." This, however, amounts to simply choosing from among the variation nature offers us. Over the past century, we have begun to go further: to actively intervene in and redesign living systems. As early as 1912 the German marine biologist Jacques Loeb wrote: "The idea is now hovering before me that man himself can act as a creator, even in living nature … Man can at least succeed in a technology of living substance."

Some consider that this vision—biology as engineering, as invention—is now becoming a reality through biotechnological innovations such as genetic engineering, in which the genetic instructions that guide biological growth are rewritten to our own specifications. Since the end of the twentieth century, the discipline called synthetic biology has attempted still deeper interventions, radically transforming the functions and forms of engineered organisms. Researchers have retooled bacteria to flash on and off with light as they generate pulses of fluorescent protein, or yeast to produce antimalarial drugs.

The ultimate expression of such goals would be the creation of a totally synthetic organism, designed on the drawing board and built in the lab. That is what a team at the J. Craig Venter Institutes (JCVIs) in Rockville, Maryland, and La Jolla, California—the research labs of a not-for-profit genomics organization founded and run by American biotechnology entrepreneur Craig Venter—reported in 2010: as the researchers themselves claimed, their experiments had produced "the first synthetic self-replicating bacterial cell."

In fact, this organism was far from entirely lab-made. It consisted of cells of the microbe *Mycoplasma mycoides* (a bacterial parasite) in which the entire genome had been synthesized artificially. The JCVI researchers, led by biologists Hamilton Smith, Clyde Hutchison, John Glass, and Dan Gibson, based this synthetic genome on that of the natural bacterium: a circular strand of DNA containing 1.2 million DNA base pairs (bp), DNA's fundamental molecular building blocks. They designed a stripped-down version just 1.08 million bp long, in which some genes considered non-essential for the organism to survive were removed. The JCVI team said that a simpler genome should be easier to understand, adapt, and redirect to give the bacterium new capabilities, "enabl[ing] us to direct cells and organisms to perform jobs, such as creating clean water or new biofuels that natural species cannot currently do."

The "synthetic organism," dubbed *Mycoplasma mycoides JCVI-syn 1.0*, was the culmination of fifteen years of research, during which the researchers developed the techniques needed to make very long strands of artificial DNA and to transfer them between organisms. The researchers constructed the synthetic genome in 1,078 segments, each 1,080 bp long, synthesized chemically by a DNA synthesis company connected to JCVI called Blue Heron Technology. The segments were assembled in yeast cells, which possess natural enzymes that stitch DNA molecules securely together. The completed genome was then transferred to a bacterium called *M. capricolum*, whose original genome was then either destroyed by DNA-snipping enzymes or just jettisoned as the cells divided. The "synthetic cells" seemed to function perfectly well with this new, streamlined genome.

The researchers inserted into the genome of *M. mycoides JCVI-syn 1.0* wholly artificial DNA segments that acted as a kind of watermark of their manufactured origin. Using a special code to translate the alphabet into DNA bases, they wrote into the genome the names of all forty-six researchers involved in the work,

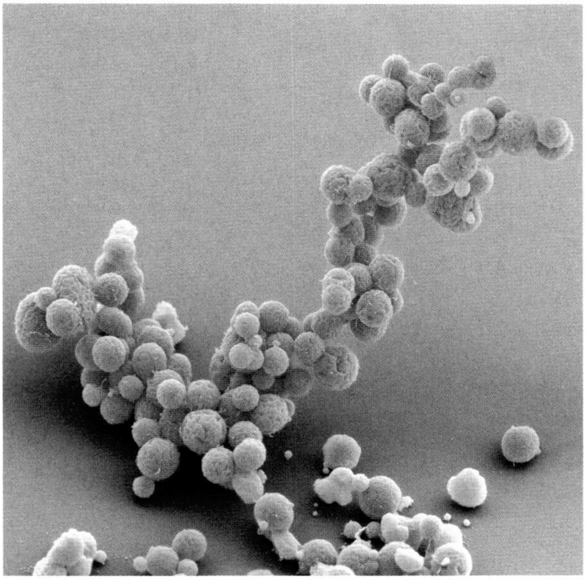

Micrograph taken by a scanning electron microscope of *Mycoplasma mycoides JCVI-syn 3.0*, a synthetic bacterium.

a website address, and a quote from the American physicist Richard Feynman: "What I cannot create, I do not understand."

Since the work required natural cells, with all their molecular machinery, to act as the host for the artificial genome, it's debatable whether *M. mycoides JCVI-syn 1.0* can truly be regarded a "synthetic organism." All the same, the implications of the experiment seemed so profound that the researchers had submitted their earlier plans to make a "minimal genome" to an independent ethics panel, which decided the work could go ahead. To ensure that synthetic organisms like these don't wreak havoc with microbiological ecosystems if they leak from the lab, the researchers said they could design them to depend on some nutrient not available in the wild, or even to carry "suicide genes" that kick in outside the controlled environment of the lab.

CRAIG VENTER | B. 1946

Perhaps stereotypically for a billionaire entrepreneur, Craig Venter was a poor student, preferring surfing to class. Only after serving in a field hospital during the Vietnam War did he begin serious education, graduating in biochemistry from the University of California at San Diego (UCSD) in 1972. He joined the US National Institutes of Health in 1984, where he learned techniques for identifying genes. He raised venture capital to found the gene-sequencing company Celera Genomics in 1998, which became involved in a race against the publicly funded Human Genome Project to sequence the entire human genome. A firm believer in the possibility of harnessing genetic engineering to solve problems such as the production of biofuels and medicines, Venter founded the J. Craig Venter Institute, with branches in Maryland and California, to pioneer synthetic biology. Venter sold the Californian facility to UCSD in 2022. Venter also cofounded Human Longevity Inc., a company seeking ways to extend the human lifespan.

See also: Experiment 49, The proof that DNA is the genetic material, 1951–1952 (page 197); Experiment 53, The cloning of Dolly the Sheep, 1996 (page 211).

2

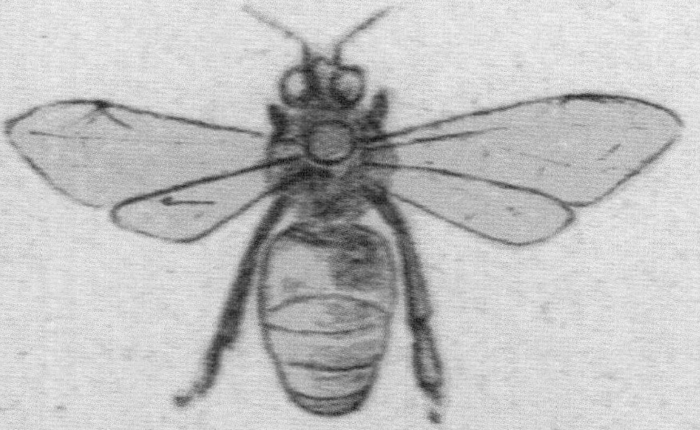

3

4

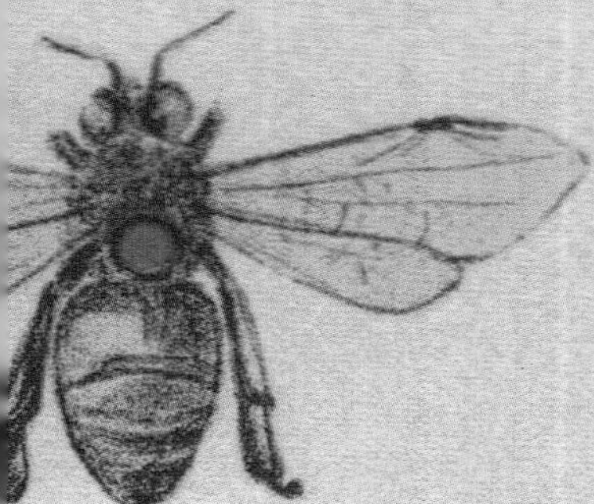

7

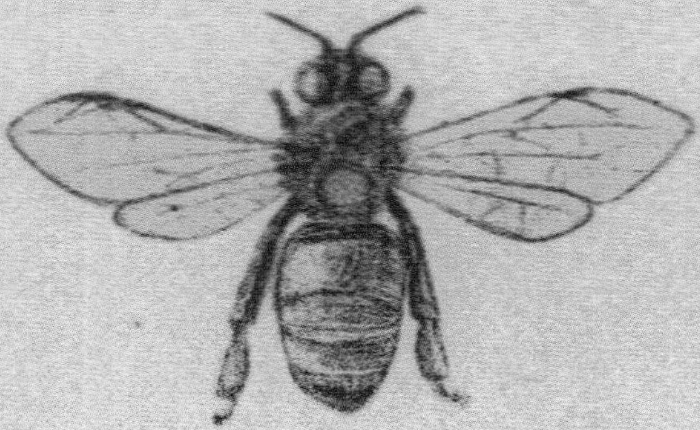

8

9

30

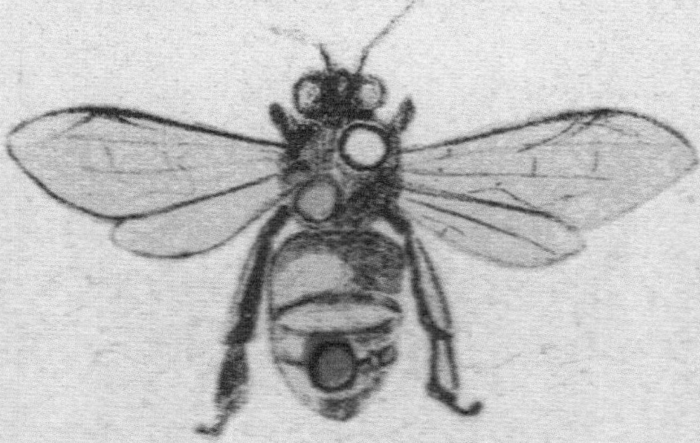

291

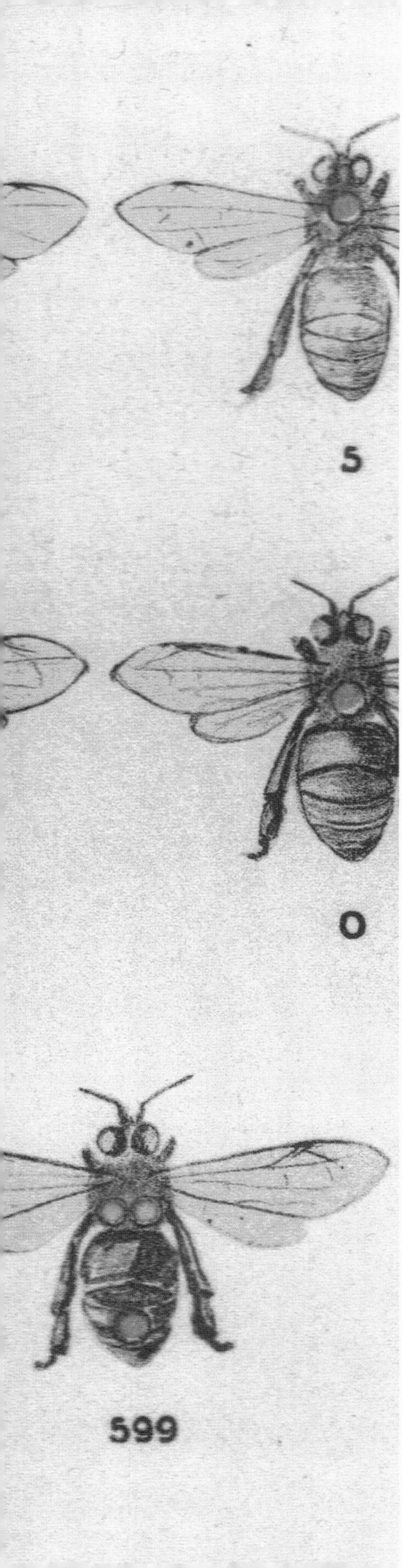

CHAPTER SIX

How do organisms behave?

The demonstration of electricity from electric fish (1772-1776)

Q Can some animals really produce electric discharges?

When the German explorer and naturalist Alexander von Humboldt was traveling in Venezuela in 1800, he was delighted to discover that shallow pools around the town of Calabozo were infested with electric eels, as he had long wished to study these peculiar South American fish. The problem was that the shock they delivered could allegedly kill a person. To solve that dilemma, the locals rounded up wild horses and forced them into a pond, where they endured terrible shocks until the fish had all but exhausted their ability to produce them. Even then, von Humboldt still received nasty jolts as he tested and dissected the eels.

The South American electric eels, such as the so-called "eel of Surinam," had fascinated European naturalists since, in the mid-eighteenth century, they had been added to the small roster of electric fish known since antiquity. These

A horse brought down by electric eel shocks. From Emil Du Bois-Reymond's *Untersuchungen über thierische Elektricität*, Berlin: Verlag von G. Reimer, 1849, Vol. 2, ETH-Bibliothek Zürich, Switzerland.

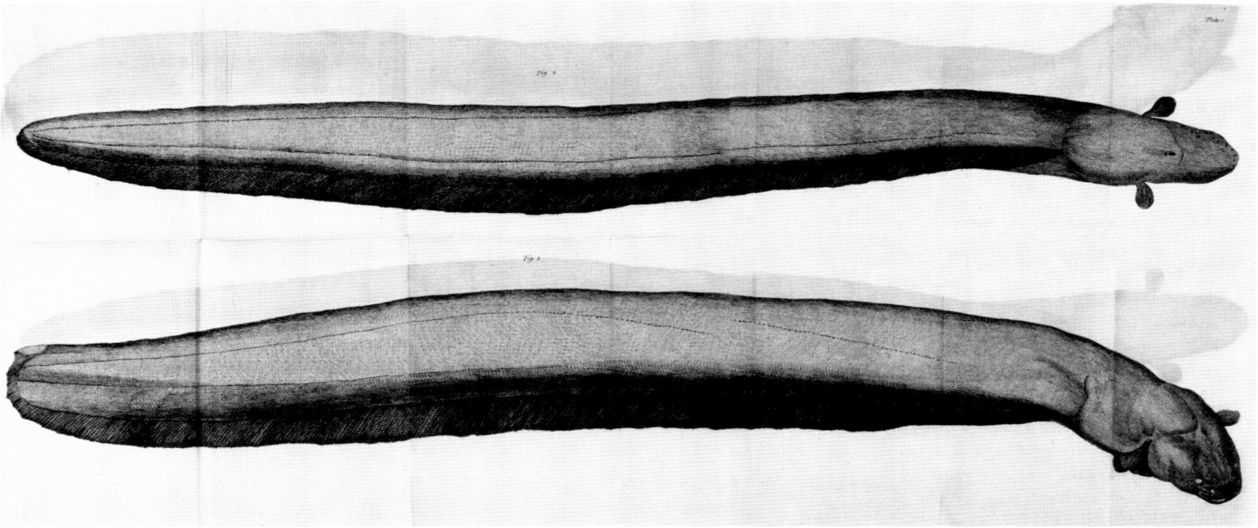

Illustration of electric eels by James Roberts. From John Hunter's "An Account of the Gymnotus Electricus," *Philosophical Transactions*, London: Royal Society, 1775, Vol. 65, Plate 1, Natural History Museum Library, London.

included the Nile catfish, portrayed in Egyptian bas-reliefs since the third millennium BC, and the torpedo rays found in the Mediterranean and coastal regions of southern Europe. The shocking effects of these animals are mentioned by Plato and Aristotle; Aristotle's successor Theophrastus reports that they can stun a person through water even without direct contact. (The name "torpedo" derives from the numbness or "torpor" the animals can cause.) The Greco-Roman doctor Galen used torpedos to deliver shock cures; the eleventh-century Persian physician Avicenna (Ibn Sina) advocated them as a headache cure.

The Italian physician Stefano Lorenzini suggested in 1678 that torpedos emit a spray of particles that penetrate the tissues and jolt the nerves, while the French naturalist René-Antoine Ferchault de Réaumur proposed in 1714 that the jolt came from the impact of an imperceptibly rapid dilation of the fish's body. But these "mechanical" explanations were challenged in the mid-century by the French botanist Michel Adanson, who said that the shock of the African electric catfish "did not appear to me sensibly different from the electric commotion of the Leyden experiment that I had tried many times"—that is, the discharge of a Leyden jar, recently invented to store electricity. That notion was supported by American doctor Edward Bancroft's observations in 1769 of an equatorial American electric eel that "communicates a shock perfectly resembling that of Electricity." Like electricity, the shock could be conducted through metal or water—indeed, fishermen had long attested that they would feel the torpedo's shocks through wet netting or along fishing lines. It could be felt, too, by a chain of people holding hands, passing down the line from one to the other.

Some felt it implausible that living creatures could be a source of electricity. How might they be expected to accumulate and retain it while immersed in a conductive liquid and made of moist tissues? While the shocks might feel similar, no one had shown that electric fish could, like a Leyden jar, discharge an actual spark. Others wondered if there might be two different *kinds* of electricity, or perhaps—as the English scientist

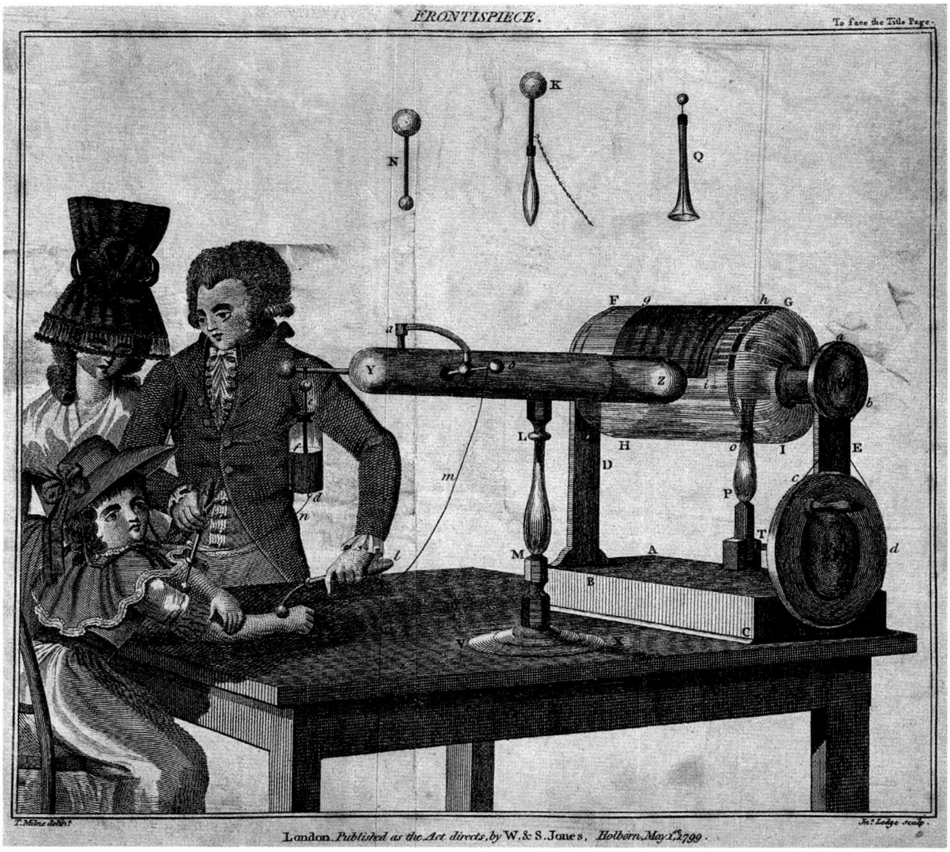

Line engraving by J. Lodge after T. Milne, showing George Adams demonstrating an electrotherapy machine with a Leyden jar for storing electricity to a woman and her daughter. From George Adams's *An Essay on Electricity*, London: W & S Jones, 1799, Wellcome Collection, London.

Henry Cavendish suggested—the "electrical fluid" in the fishes might be less dense than that in the jars. Cavendish showed that the shock delivered by the fish was more like that produced by many linked and weakly charged Leyden jars, rather than a single, highly charged one that caused a spark.

The English naturalist John Walsh, a Fellow of the Royal Society in London, took it upon himself to resolve the matter. After corresponding on the subject with American scientist Benjamin Franklin, perhaps the foremost expert on electricity at the time, in 1772 Walsh traveled to La Rochelle on the French coast to study the torpedo. There he experienced his first shock from one of the rays and verified that it could be transmitted through metal and from person to person, but that sealing wax blocked it. Walsh spoke of an "electric organ" in the fish that generated the discharge. The British anatomist John Hunter dissected the torpedo and showed that the organ contained many stacked membranes: seemingly an inspiration for the stacked metal plates of the "voltaic pile" devised by Alessandro Volta, who corresponded with Walsh.

Walsh's findings, which he published in 1773, won him the Royal Society's distinguished Copley Medal the following year. But still he sought that elusive spark from a fish: clinching proof that electricity was at the root of it.

He finally succeeded with a Surinam eel imported from Guiana, on which he experimented

The electric eel, *Electrophorus electricus* (formerly *Gymnotus electricus*), Smithsonian's National Zoo & Conservation Biology Institute, Washington, DC.

in his London house. These eels produce a voltage about ten times greater than that of torpedos, and in 1776 Walsh finally managed to draw from the creature a spark that traveled across a narrow gap in a tin strip fastened to the glass holding the fish. The experiment was conducted in the dark so that the tiny spark was easier to see. Walsh brought friends and colleagues to his house to witness the effect, and found that the spark created a shock that could be felt throughout a chain of twenty-seven people.

The finding was incontestable, persuading even some who had previously expressed scepticism about "animal electricity." Walsh described the experiment in a letter to the French naturalist Jean-Baptiste Le Roy, who translated and published it in a French journal.

This was the closest that Walsh ever came to actually publishing his result; in those days, testimony of expert witnesses was still sometimes regarded as evidence enough. It is perhaps for this reason that, although the findings were widely known at the time, they were rather rapidly forgotten. When Michael Faraday repeated some of Walsh's experiments on electric fish in 1839, he expressed doubts about whether Walsh had conducted them after all. Nonetheless, Walsh's demonstration that animals could be autonomous sources of electricity motivated the studies of Luigi Galvani on electrical stimulation of nerves (see page 172), which launched the field of electrophysiology.

JOHN WALSH | 1726–1795

John Walsh was born in colonial India, and as an employee of the East India Company from 1742 to 1757 and private secretary to Lord Clive (whose wife was the niece of Walsh's mother), he accumulated a considerable fortune there. This made him a gentleman of leisure, from which position he could indulge his passion for natural history. Sent to England by Clive to help coordinate plans for reorganizing the government of Bengal, he served as a Member of Parliament from 1761 to 1780 and was appointed a Fellow of the Royal Society in 1770.

See also: Experiment 16, The discovery of alkali metals by electrolysis, 1807 (page 72); Experiment 42, Animal electricity, 1780–1790 (page 172).

What the earthworm knows (1870-1880s)

Q Can we see intelligence even in "lower animals"?

The mere fact that Charles Darwin's last book was *The Formation of Vegetable Mould, Through the Action of Worms*, published in 1881, speaks volumes about the man's character. Having revolutionized all of biology with his *Origin of Species* (1859) and then emphasized in *The Descent of Man* (1871) how his ideas about evolution transformed the understanding of our own position in the world, one might have imagined Darwin settling in later life into the role of grand speculative philosopher. Instead he stayed true to his passion for the minutiae of the living world by studying this most humble of topics: how soil is formed by worms. There was perhaps something in the modest but immensely important toil of these creatures with which Darwin identified. "It may be doubted," he wrote, "if there are any other animals which have played such an important part in the history of the world as these lowly organized creatures."

Print after a daguerreotype of Charles Darwin with his son William Erasmus, who later assisted in his experiments on earthworms, 1842, Cambridge University Library.

Despite—or is it because of?—the apparently prosaic subject matter, Darwin's book on mold and worms has tremendous charm. It is no sweeping synthesis in the manner of the *Origin*, but rather, based on careful observation and experiment, much of it conducted in Darwin's own garden with the help of his family members. He and his children would set out early each morning from their country house at Down in Kent to examine the work of the earthworms in the cool, moist soil. These creatures munched through leaves and other vegetable matter, turning it in their gut into rich, fertile hummus. Their individual labors might seem trivial, but Darwin calculated that over a year the worms might move 8 tons of earth in every acre of land, and their actions could eventually bury entire ancient buildings and monuments. He was evidently impressed by what slow, almost invisible but steady labor can achieve, given time.

The Formation of Vegetable Mould is filled with descriptions of experiments on worms and their habits, sometimes of delightful eccentricity—as when Darwin placed the creatures in pots of earth on top of his piano and watched their responses to the sounding of different notes ("C in the bass clef" and "G above the line in the treble clef" cause them to retreat into their burrows, though presumably

Earthworm stone with a central measuring peg, used by Darwin for his garden experiments at Down House, Kent, Wellcome Collection, London.

not uniquely), or investigated how they responded to his breath when he chewed tobacco or placed drops of vinegar in his mouth. But it is Darwin's studies of the burrow-making activities in the wild that testify most to his boundless curiosity and inventive experimental style. Darwin wondered about the principles governing the way worms plug the openings of their burrows with leaves—an activity they perform with such vigor, he says, that at times the rustling sound may be heard on still nights. Whether they do it for protection from predators or rainwater or cold air, or for food, he admits to being unsure; maybe a bit of all of them.

What fascinated Darwin most about this behavior was the evidence that it supplied of the worms' intelligence. "If a man had to plug up a small cylindrical hole, with such objects as leaves, petioles [stalks] or twigs," he wrote, "he would drag or push them in by their pointed ends; but if these objects were very thin relatively to the size of the hole, he would probably insert some by their thicker or broader ends." Do worms likewise adapt their strategy to the objects at hand, or do they just perform their actions at random? Darwin reported that 80 percent of leaves he removed from worm burrows had been inserted tip first—a far from random distribution. He then proceeded to consider the differences for leaves of various species and shape. After nighttime observations in dim light with his son Francis (who became a distinguished botanist in his own right), Darwin

CHARLES DARWIN | 1809–1882

Many famous scientists are admired for their works, but very few enjoy the reverence afforded to Charles Darwin. It doubtless helps that his life story has adventure: the formative voyage around the world aboard the HMS *Beagle* from 1831–36, personal tragedy (the death of his beloved daughter, Annie, in 1851), and a crisis of faith as his scientific theory of evolution by natural selection raised questions about the nature of man and the need for a God. His appeal is surely boosted, too, by his modesty—he was scrupulous in acknowledging that his theory of natural selection was much the same as that of Alfred Russel Wallace, even if it was Wallace's manuscript in 1856 that spurred Darwin finally to commit his own ideas to paper. And there is a splendid bathos in the way that Darwin was delayed in announcing probably the most important idea in all of biology not just by his personal doubts and lack of confidence but also by his commitment to writing several monographs on barnacles. He was arguably a botanist at heart, writing books on orchids, climbing plants, insectivorous plants, plant fertilization, how plants move, and, finally, on the way dead plants are transformed to soil. Those studies meant at least as much to their author as the theory of how all life on earth came to be the way it is.

See also: Experiment 57, Behavioral conditioning as window to the mind, 1903–1936 (page 226); Experiment 59, Decoding the waggle dance of bees, 1919–1940s (page 230).

attested that "It appeared to both my son and myself as if the worms instantly perceived as soon as they had seized a leaf in the proper [that is, the most efficient] manner."

Not content with these studies using natural leaves, Darwin made a more controlled and systematic investigation of the effects of shape on the worms' behavior using cut-out paper triangles of varying proportions, "rubbed with raw fat on both sides" to protect them from damp. "We may infer," he decided, "that worms are able by some means to judge which is the best end by which to drag triangles of paper into their burrows."

This suggested to Darwin that the humble earthworm is no mere automaton guided purely by instinct. "We can hardly escape from the conclusion that worms show some degree of intelligence in their manner of plugging up their burrows," he wrote. "It is surprising that an animal so low in the scale as a worm should have the capacity for acting in this manner."

In truth, Darwin's reasoning might be criticized as anecdotal. He contrasts the worm's intelligence

Diagram of the alimentary canal of an earthworm (*Lumbricus*). Drawing annotated by Charles Darwin after Ray Lankester, *Quarterly Journal of Microscopical Science*, Vol. XV, N.S. Plate VII, Cambridge University Library.

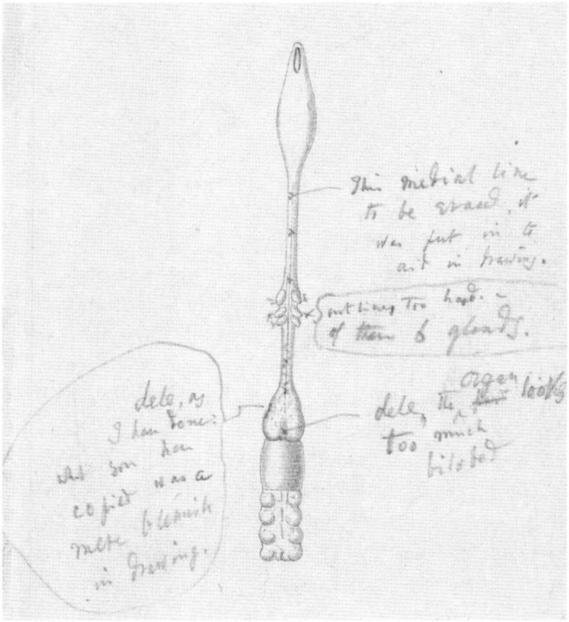

Linley Sambourne's illustration "Man Is But A Worm," a caricature of Charles Darwin. It appeared in the magazine *Punch* on December 6, 1881, the year Darwin published his last book *The Formation of Vegetable Mould, Through the Action of Worms*, *Punch Almanack*, London: At the Office, Fleet Street, 1882, Heidelberg University Library, Germany.

with the stupidity of the bee that buzzes for hours trying to escape through the wrong side of a half-opened window, but today there is abundant evidence of the ability of bees to learn, communicate, and perhaps even have a kind of variable mood inflected by experience. Yet his basic methodology for studying behavior is sound: to create experimental situations in which advantageous and apparently rational choices of action may be distinguished from random ones, and to gather good statistics on the outcomes. Perhaps most commendably of all in an age when human exceptionalism was still the norm, Darwin was refreshingly open-minded about the potential cognitive powers of "lower" animals—an attitude now reflected in the "biopsychist" belief of some biologists that a kind of feeling or sentience obtains in all living things.

Darwin's book was an unlikely bestseller, shifting more copies in its day than the *Origin of Species*. Perhaps that says something about the English people, whose enthusiasm for overarching theories of nature pales before their passion for gardening.

Behavioral conditioning as window to the mind (1903–1936)

Q What kind of interior world do animals have?

According to Aristotle, humans are elevated above animals by possessing a "rational soul" that gives us powers of reason. Animals, in contrast, have only a sensitive soul, so that they can feel stimuli and respond to them only in the manner of automata. For René Descartes in the seventeenth century, animals were every bit as mechanistic as the physical universe: complex machines best understood as systems of pumps, levers, and hydraulics. (Descartes' view of the human body was little different, and the mere fact that he awarded us also, and uniquely, a soul, did not save him from accusations of heresy and atheism.) While such a reductive view of animal sentience prevailed, it was hardly surprising that natural historians felt warranted in conducting sometimes grotesque vivisection experiments on live animals.

This rather brutal approach to animal research persisted through the nineteenth century, exemplified in the experiments conducted by the Russian physiologist Ivan Pavlov, head of the department of physiology at the Imperial Institute of Experimental Medicine in St. Petersburg. To Pavlov, an organism was a kind of biological machine governed by rules dictating the action of its component parts, and the goal of the physiologist was to elucidate those rules. His work focused on digestion, and his laboratory animal of choice was the dog.

It is not easy to reconcile this disconnected attitude with Pavlov's evident affection for dogs. He had his favorites among those bred in his lab, and never doubted that dogs are conscious beings with feelings and emotions, including joy. And yet he treated dogs as a kind of technology, instructing his assistants to operate on live animals as if to transform them into instruments with which he could measure and quantify isolated aspects of digestive processes. Pavlov insisted that such experiments should embody the objective scientific spirit, "dealing only with external phenomena and their relations"—with what

IVAN PAVLOV | 1849–1936

Ivan Petrovich Pavlov trained initially for the priesthood before enrolling in 1870 at the University of St. Petersburg to study natural sciences. There he became a protégé of the great Russian physiologist Ilya Fadeevich Tsion. After an itinerant early career, he finally settled at the newly inaugurated Institute of Experimental Medicine in 1891, where he secured his reputation for rigorous experimental work. He is one of those scientists, like Albert Einstein, whose Nobel prize was awarded for work other than that for which they are best known: it was granted in 1904 for Pavlov's studies of digestion, at which point his studies of conditioned reflexes in dogs were only just beginning. Pavlov was a fierce critic of the persecution of intellectuals under Stalin in the final two decades of his life.

See also: Experiment 58, Innate and acquired habits, 1947–1949 (page 228); Experiment 60, Mental time travel and the concept of other minds in birds, 2001 (page 232).

Five dogs undergoing experiments on gastric secretion in the Physiology Department, Imperial Institute of Experimental Medicine, St. Petersburg, 1904, Wellcome Collection, London.

one could see and measure, rather than making any inferences about the internal "psychic" life of the creature itself.

All the same, Pavlov's extensive studies of the digestive process in the late nineteenth century led him eventually to recognize that the psyche could not be excluded from an understanding of behavior. In Pavlov's studies of salivation in dogs, he observed that this response could be triggered by the smell or the mere anticipation of food—just as it can in humans. This in itself had been long known, but over the course of three decades of research at the start of the twentieth century Pavlov turned such a "conditioned reflex" into a reliable laboratory procedure that could be used to probe the mind of his dogs.

He found dogs could learn to associate some conditional stimulus with food, if the stimulus was followed by feeding. Then the stimulus itself—light, sound, cooling of the skin, even mild electric shock—would trigger salivation even if food did not follow it, although that response would gradually fade without such a payoff. By changing the parameters of the stimulus and looking at the salivation response, Pavlov could deduce how well dogs discriminate between different stimuli—for example, concluding from experiments with metronomes that they can distinguish timing differences of 1/43 of a second. In this way, he used conditional stimuli to quantify the subjective experiences of the animals: how well they can detect changes of color, temperature, distance, and so on. So precise and revealing were his techniques that Pavlov asserted we would scarcely learn any more about their interior world if dogs could tell us about it directly.

Some misunderstood Pavlov's studies as showing only that dogs drooled in anticipation of a treat: George Bernard Shaw commented loftily that "any policeman can tell you that much about a dog." Rather, Pavlov began to show how researchers might make deductions about cognition—hidden from any direct view—from careful observations of behavior. His ultimate goal, he averred in 1903, was to "explain the mechanism and vital meaning of that which most occupies Man—our consciousness and its torments." Beyond such innovations, his work, performed by a large team of assistants in a kind of "laboratory factory," heralded the dawn of Big Science in the modern age.

Innate and acquired habits (1947-1949)

Q Are behaviors learned or instinctive?

Just as the typical scientific paper reports a sanitized and retrospectively organized version of what actually transpired both in the laboratory and the minds of the experimenters, so the history of science is apt to become a curated narrative in which ambiguities and complications become smoothed away like the rough edges of a pebble. We have a dangerous fondness for the blinding eureka moment of an experimental revelation; in reality these are vanishingly rare.

This is nowhere more true than in one of the classic studies in behavioral science, conducted in 1947 by the Dutch ethologist Nikolaas Tinbergen. It is considered to be one of the first demonstrations that animals do not learn every aspect of their behavior, as was long believed, but also have hardwired instincts shaped by evolution.

Tinbergen studied the feeding behavior of herring-gull chicks. The adult bird feeds the chicks by regurgitating half-digested food and presenting a morsel held in the tip of its beak, at which the chicks will clumsily peck. In 1937, German ornithologist Friedrich Goethe realized that the chicks direct their pecking at a red patch on the lower bill, which acts as a kind of guide toward the food. Goethe showed that herring-gull chicks will peck more often at red objects. Was this instinctual behavior, or could they learn to peck at other spots? To find out, Tinbergen presented chicks in the wild with cardboard bird heads that had spots of other colors on the beaks: red, black, blue, white, as well as the plain yellow of the beak.

The usual story is that he found the chicks would peck more often at a red spot—a behavior that had no obvious learned basis, and so was presumably innate. But in fact, his 1947 study (published the following year) showed, rather to his surprise, that the chicks pecked more often at the black spot. Tinbergen concluded that what mattered for the chicks was not the color but the contrast between the spot and the beak.

But in 1949 he argued that the earlier studies had been misleading, since they always presented

NIKOLAAS ("NIKO") TINBERGEN | 1907–1988

Tinbergen shared the 1973 Nobel prize in medicine or physiology with Karl von Frisch and Konrad Lorenz, and the trio is widely considered to have brought modern ethology—the study of animal behavior—into the mainstream, making it a precise experimental science rather than an anecdotal pursuit of amateur natural historians. Tinbergen's 1951 book *The Study of Instinct* summarizes much of his work on innate behaviors—those that do not need to be learned. He worked with birds, bees, and mammals, and also on the psychiatric disorders of humans and on child autism. After incarceration in the Netherlands during the Second World War, he moved to the University of Oxford in 1949, where his students included Richard Dawkins and Desmond Morris.

See also: Experiment 57, Behavioral conditioning as window to the mind, 1903–1936 (page 226); Experiment 59, Decoding the waggle dance of bees, 1919–1940s (page 230).

comparative pairings of a red spot and that of another color. As a result, he argued, the chicks were exposed much more to red than to other colors, and became habituated to it, making them apt to ignore it once they found it did not guarantee they would receive food. (Habituation is similar to the way we filter out inconsequential noises from our perception.) After further studies guided by this idea, Tinbergen decided that his earlier data needed "correcting" to allow for the habituation effect, whereupon they showed the expected preference for red after all. Quite quickly, however, his subsequent descriptions of those experiments neglected to mention this correction, making it sound as though red had been preferred all along. The story diverged ever more from the experiments themselves with each retelling.

It seems that in his interpretations of the data Tinbergen was guided more by intuition than by objective assessment. Yet that intuition seems to have been correct. In a study in 2009, behavioral biologist Carel ten Cate of Leiden University in the Netherlands repeated the experiments and found that, indeed, chicks innately prefer red spots but can be quickly habituated to neglect them.

There is no suggestion Tinbergen intended to mislead or that his approach was substandard. Rather, standards in science change: what was accepted practice many decades ago often would not pass muster today. There is now no doubt that much animal behavior is motivated by instinct rather than learning. And Tinbergen's simple and systematic manipulations of behavior in natural settings—he later examined the effects of changing the model head's color, shape, bill length, orientation, and so on—are also much emulated.

What's more, researchers are now more attuned to the potential to be misled by their expectations and unconscious desire for a particular result, a phenomenon known as cognitive bias. As the physicist Richard Feynman said, in science . . . "The first principle is you must not fool yourself, and you are the easiest person to fool." Experimenters often take steps to avoid fooling themselves, for example, by performing double-blind tests (so no one knows until the data are collected and analyzed which is the "test group" and which the control group) and pre-registering their hypothesis and methods so they don't end up changing the question to fit one that the data can answer. All the same, studies in behavior and psychology—especially of humans—are notoriously prone to being non-reproducible, to a degree that has alarmed these fields in recent years. But that is perhaps not surprising. When it comes to behavior, it is hard to devise and conduct an experiment in the classic manner of science, leaving all but one variable unchanged. We, and other animals, seem sometimes to be extremely sensitive to context and to small differences.

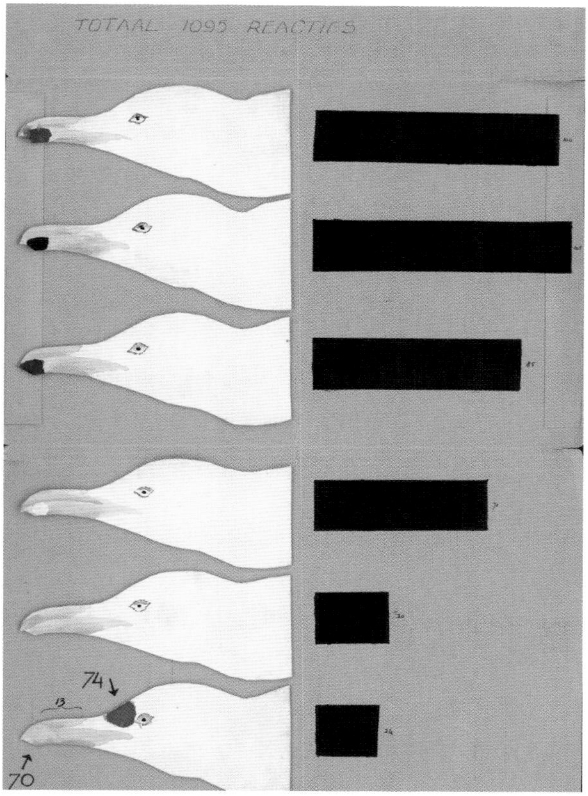

Niko Tinbergen's diagram showing chick pecking responses to herring gull beaks marked with different colors, Rijksmuseum Boerhaave, the Netherlands.

Decoding the waggle dance of bees (1919-1940s)

Q How do bees convey information while foraging?

The more we understand the cognition of other animals, the less special the human mind seems. It is now widely accepted that at least some other species have a degree of consciousness, emotions, and moods, complex social interactions, and a sense of past and future. But in at least one regard we still seem to be unique: we are the only known species to have a genuine language, as distinct from a simple repertoire of calls and vocalizations that signify alarm, mating signals, and the like. Human language possesses a syntax relating word order to meaning, and—crucially for cultural learning—it can be expressed in the abstract symbolic form of writing. But symbolic communication in itself is not solely the human preserve it was once thought to be. There is at least one other animal that communicates by using symbols arbitrarily to represent information: the honeybee.

When a bee finds a food source while foraging, it returns to the hive and tells the other bees where it can be found, so fellow hive members may also locate it, for the good of the colony. This information is conveyed solely by the bee's movements. For sources close by (within about 300 feet), it is indicated in a "round dance" in which the bee moves on the comb in narrow circles clockwise and counterclockwise. For more distant sources, the communication is more complex, employing the so-called waggle dance. The bee begins by moving in a straight line while waggling its posterior. Then it circles around back to the starting point and repeats the motion, first to the left and then the right in a figure of eight. "The waggle dance looks comical," the German-Austrian ethologist Karl von Frisch attested. "But it is not really comical, it is incredibly interesting. It is one of the most amazing occurrences in the insect world."

Von Frisch began studying bees when he was first appointed a professor of zoology at the University of Munich in 1912. His initial studies set out to establish that, contrary to prevailing belief at the time, bees have color vision. From 1919 he began to investigate how bees forage, making the crucial discovery that they could be trained to locate food at artificial feedings stations, so one could control where they went. He noticed that bees make changes in their dance depending on the direction and distance of the food stations, and figured that the movements were conveying specific information. Von Frisch devised a system for marking bees with dots of colored lacquer, so he could keep track of individuals in a swarm

Hundreds of bees marked with tiny dabs of paint, to identify and observe their individual movements. From Karl von Frisch's *Über die "Sprache" der Bienen*, Jena: G. Fischer, 1923.

(even, with practice, while in flight), and he used the latest technologies, especially photography and film, to gather his data. His studies of foraging grew into an enormous body of experiments conducted over three decades.

Von Frisch deduced that the waggle dance encodes in symbolic form two pieces of information to guide other bees to the food source. The direction of the central linear portion of the dance shows the direction from the hive: amazingly, not as a simple pointer, but in coded form. The angle of this line relative to the vertical orientation of the comb shows the angle the bees must fly relative to the line from the hive to the Sun—strictly speaking, to the azimuth, the direction of the horizon directly beneath the Sun. Von Frisch positioned feeding stations in fanlike arrays around the hive to deduce how precise this instruction (and its execution) was.

The duration of the dance meanwhile signifies the distance to the destination, roughly speaking with each second of the dance denoting a kilometer of travel. Von Frisch also discerned local variations in this code for different bee species, rather like dialects, and that they may compensate in their instructions for changes in flight time due to winds. "Equivalent accomplishments are not known among other animals," he claimed.

The information in the dance is not accurate enough to precisely specify the food source, but bees can find it by smell once in the vicinity. Some researchers, especially American biologist Adrian Wenner, doubted von Frisch's conclusions, arguing that the bees navigate by smell alone—and, indeed, Wenner noted in the 1960s that a waggle dance won't work for food with no odor. (Foragers carry the odor of their food source back with them via waxy hairs on their legs.) Even today some debate continues about the relative importance of the two sources of information, although strong evidence for the veracity of von Frisch's interpretation of the waggle dance was reported in 2005 from radar measurements of bee trajectories.

Bees are now known to possess some impressive cognitive abilities, including learning complex tasks (like pushing a small ball into a hole) for rewards. Their ability to construct geometrically precise hexagonal combs of wax has long fascinated natural philosophers, leading some in the eighteenth century to suppose they were instructed in the rudiments of geometry "by divine guidance and command." As the American ethologist James Gould, who confirmed von Frisch's work on the waggle dance, said, "There is perhaps a feeling of incongruity in that the honey-bee language is symbolic and abstract, and, in terms of information capacity at least, second only to human language. Despite expectations, however, animals continue to be more complex than had been thought."

KARL VON FRISCH | 1886–1982

Born in Vienna, Karl von Frisch later proclaimed that "I was born with a love of the animal world." As a child he had not so much pets as a veritable zoo of 170 different species: "animals of all kinds were permanent guests in my nursery," he said. His father wanted him to be a doctor, but he switched his studies from medicine to zoology and became a lecturer in the subject at the University of Munich in 1912. During the Nazi era, the fact that von Frisch had some Jewish grandparentage marked him as a "non-Aryan" and so he faced possible expulsion from his university position. It was only because his work on bees, known to be vital crop pollinators, was seen as a potential help against the *Nosema* fungus that threatened them, and thus the German food supply, that he kept his post. In 1973 his work on the waggle dance of bees won him the Nobel prize in medicine or physiology along with Nikolaas Tinbergen and Konrad Lorenz—the first to be given for animal behavioral science.

See also: Experiment 56, What the earthworm knows, 1870–1880s (page 222); Experiment 58, Innate and acquired habits, 1947–1949 (page 228).

Mental time travel and the concept of other minds in birds (2001)

Q Do birds have a notion of the past and the future?

The history of animal behaviorism has long been a story of human exceptionalism. Despite Darwin's insistence on continuity between humans and other animals—there are, he claimed, "no fundamental differences between man and the higher mammals in terms of mental facilities"—there has remained a reluctance to acknowledge any flicker of consciousness, feeling, and emotion in our non-human kin.

Some of this hesitation is understandable, even warranted. There is a strong temptation to assume that any behavior resembling our own is guided by a comparable state of mind, so that we can end up unwarrantedly anthropomorphizing other animals. Some animal behaviorists now argue that we are

"The Crow and the Pitcher." A wood engraving from Thomas Bewick's *Bewick's Select Fables of Aesop and Others*, London: Bickers, 1871, Part II, University of Toronto – Robarts Library.

A New Caledonian crow taking part in an experiment to test whether, given the choice between using a solid or hollow object, it would choose the solid object to displace the water in the cylinder in order to access a floating piece of food. All the birds that participated correctly chose the solid object nearly 90 percent of the time.

prone to over-compensating, finding convoluted reasons why an animal did what it did through blind and automatic instinct when the most parsimonious explanation would be to assume some similarity with our own sensitive minds.

Teasing out just what other animals do and don't think or feel requires tremendous ingenuity, and only in the past several decades have such studies started to reveal with confidence that we may be less unique than we have supposed. It is perhaps not so surprising to find similarities with the minds of other primates, such as chimpanzees. But it is more challenging to understand the cognitive processes of animals whose physiology, anatomy, and indeed goals have less in common, such as birds.

Birds are very diverse in their cognitive abilities. Some are able to sense the Earth's magnetic field for navigation; some voice complex but reproducible songs as mating calls; many

Professor Nicola Clayton with some rooks at the Comparative Cognition Laboratory that she set up at Cambridge University.

NICOLA CLAYTON | B. 1962

Nicola Clayton graduated in zoology from Oxford University and studied birdsong for her doctorate at the University of St. Andrews. She set up her Comparative Cognition Laboratory at the University of Cambridge in 2000 to study the behavior of corvids in particular. It is long-term work that requires establishing a relationship with the birds. "It's a privilege," says Clayton, "to get the opportunity to see inside their minds, and for them to trust us enough to share what they know with us." This need for sustained investment in such research made it alarming when, in 2021, Clayton's laboratory faced closure after its funding was not renewed. After outcry from other academics, a public appeal has raised £500,000 to keep it open. Clayton is also an accomplished dancer, and since 2011 has been Scientist in Residence with the Rambert Dance Company.

See also: Experiment 56, What the earthworm knows, 1870–1880s (page 222); Experiment 58, Innate and acquired habits, 1947–1949 (page 228).

construct elaborate nests, and the bower bird seems to show something that looks like aesthetic judgment in building (for males) or evaluating (for females) the "bowers" that feature in their mating ritual. Some, such as corvids, use tools; others, like parrots, are excellent sonic mimics.

The question remains, however, of whether these skills are conducted with any glimmer of consciousness and self-awareness. One of the central questions for animal behaviorists is whether other animals possess a so-called Theory of Mind, meaning that the creature sees others as separate entities with their own motives and perspective—for example, whether they appreciate that they themselves can have knowledge that others do not possess, and vice versa. Another question is whether animals have any notion of past and future: it was long thought that this "mental time travel" is solely a human attribute and that other animals live in a perpetual present.

Both of these qualities of mind were studied in an experiment in 2001 by animal psychologist Nicola Clayton of the University of Cambridge and her husband, cognitive scientist Nathan Emery. They studied Florida scrub jays, a species of corvid native to North America, which store food in the wild as caches to which they can return. The caching is fairly unsophisticated—it's rather easy to spot the stored food—and so pilfering of one bird's cache by another is common. In particular, scrub jays that watch another bird cache food may return to that location to sneakily steal it. But the birds might also try to foil such thievery by returning to their own caches and moving them elsewhere.

Clayton and Emery set up an experiment to study how the caching strategies of individual birds vary. They provided the birds with sand-filled trays for caching worms, and allowed them to return to their caches after three hours, when the birds could feed but also had the option of re-caching uneaten food; the researchers provided a second sand tray as an option for that. They found that the scrub jays recached significantly more often if they had been observed by another bird—and were aware of it—during initial caching. What's more, they recached more often in the new tray, presumably deciding it was a safer place.

This behavior in itself suggests the birds are aware that a cache is more at risk of pilfering if another bird has watched them store it—as if they are thinking "he saw me do it, so he knows it's here." That in itself doesn't tell us what really goes on in the bird mind, however: perhaps it simply has a hardwired instinct to "recache if observed." But Clayton and Emery saw something more. They conducted further experiments using a group of birds to which they had previously given an opportunity to pilfer from others: these birds, as it were, "had form" in thieving. They

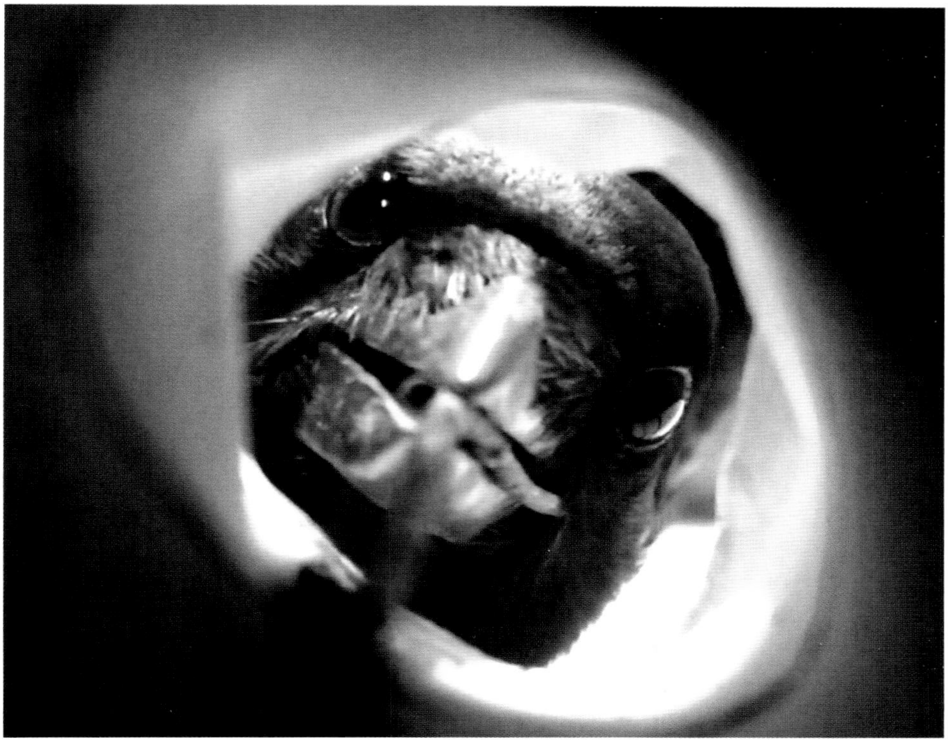

Video still showing a crow probing at larvae with a twig in an experiment conducted in 2012, courtesy of Dr. Jolyon Troscianko of the University of Exeter.

found that the experienced pilferers, if observed while caching, recached more items than a group that had not had this training. Now it is as if the birds think "Well, I know *I've* stolen before, so I figure others will too." The birds seem able to attribute to others the potential for a motive or behavior they have experienced themselves in the past. As Clayton and Emery wrote, "this is the first experimental demonstration that a non-human animal can remember the social context of specific past events, and adjust their present behavior to avoid potentially detrimental consequences in the future." In other words, the birds showed both mental attribution—a kind of theory of mind—and mental time travel.

The art in animal studies of this sort is to devise experiments that are controlled enough to draw reliable conclusions while not making the situation so artificial that it can't reliably show us what the animal might do in the wild. Often, the natural behavior might be consistent with a variety of interpretations, and the aim is to narrow the options so that we can conclude something reliable about the inaccessible mind producing it.

Further reading

As far as was possible, I have consulted the original sources for all of the experiments described in this book. This is often less daunting than it might sound: the works of Humphry Davy and Michael Faraday, for example, are delightfully engaging and lucid. It is often eye-opening too: retrospective accounts tend to clean up the story, sometimes in ways that do not reflect what was really done.

The idea of surveying the history of experimental science by focusing on paradigmatic experiments is not new. It was done by Rom Harré in *Great Scientific Experiments* (Phaidon, 1981), which includes a more detailed discussion than I have space for of the philosophy and methodology of experimentation. A smaller selection of experiments, with a strong emphasis on physics, was presented in Robert Crease's *The Prism and the Pendulum: The Ten Most Beautiful Experiments in Science* (Random House, 2004). Crease's book explores the aesthetic element of experimentation with great subtlety. My own book *Elegant Solutions: Ten Beautiful Experiments in Chemistry* (Royal Society of Chemistry, 2005) was both homage and riposte to Crease, making the case that chemistry too has plenty that is admirable and even beautiful in its experimental history. George Johnson offered a complementary selection in *The Ten Most Beautiful Experiments* (Bodley Head, 2008).

The challenge in all these exercises is the same one faced in this book: that by focusing on specific experiments, one can give the false impression that science advances in a succession of eureka discoveries, often made by lone (and moveover Western and male) scientists using a standard set of methodological tools. The need to examine specific experiments in depth also tends to produce a focus on developments since the early modern period (from around the seventeenth century), since records from earlier times are often very sketchy. My hope is that the accounts in this book reveal this to be at best a partial view of how science works. The classic account of how early modern science (in particular the "experimental philosophy") was socially mediated is Steven Shapin and Simon Schaffer's *Leviathan and the Air-Pump* (Princeton University Press, 1985); an alternative view of that era is offered in David Wootton's *The Invention of Science* (Allen Lane, 2015). And an excellent account of the messy realities of doing modern science is given in Jeremy Baumberg's *The Secret Life of Science* (Princeton University Press, 2018). The relatively recent genesis of the alleged "scientific method," and its rhetorical functions, is explored in Henry Cowles's *The Scientific Method* (Harvard University Press, 2020). On the philosophy of experimental science there is an immense literature, to which Pierre Duhem's *The Aim and Structure of Physical Theory* (Princeton University Press, 1991) was an influential early contribution. You should not suppose there is any consensus today on what "experiment" means!

For the long view of experimental science, Lynn Thorndike's monumental multi-volume work *A History of Magic and Experimental Science* (Columbia University Press, 1923–41) remains an invaluable source, although a great deal of detailed scholarship has augmented and modified that history since. Much of that scholarship informs *The Cambridge History of Science*, Volumes I (2018) and II (2013), which offer detailed accounts of, respectively, ancient and medieval science, including the fragmentary record of what experiment might mean in those times—not just in classical civilizations and the West but also in, for example, China, India, and Babylonia. Paul Keyser and John Scarborough's *The Oxford Handbook of Science and Medicine in the Classical World* (Oxford University Press, 2018) does a similarly excellent and more concise job of that. While the nature of this book, with expositions on individual experiments, militates against close investigation of the experimental practices of times before detailed records exist, that should not be taken to imply that humans have not, since time immemorial, engaged in what we might regard as an experimental manner with the world around them.

Finally, the aesthetic aspects of science have received far less attention, until recently, than is warranted. *The Aesthetics of Science: Beauty, Imagination and Understanding*, edited by Milena Ivanona and Steven French (Routledge, 2022), makes an excellent start to correcting that oversight. Astrophysicist Subrahmanyan Chandrasekhar's *Truth and Beauty: Aesthetics and Motivations in Science* (University of Chicago Press, 1987) offered a thoughtful personal view of the issues. That they are now being given serious scholarly attention is exciting and overdue.

Index

Page numbers in *italics* refer to illustration captions.

Credits

Picture credits

Every attempt has been made to trace the copyright holders of the works reproduced, and the publishers regret any unintentional oversights. Illustrations on the following pages are generously provided courtesy of their owners, their licensors, or the holding institutions as below:

Adilnor Collection, Courtesy the: 140; Akira Tonomura, from John Steeds et al 2003, *Phys. World 16* (5) 20, fig. 2: 164 (bottom); Alamy: 65 (The Natural History Museum); 88 (Science History Images): 156 (bottom; Pictorial Press Ltd); 175 (Chronicle); 213 (Universal Images Group North America LLC); ATLAS Experiment © 2022 CERN: 133; Bavarian Academy of Sciences https://publikationen.badw.de/en/003395746 (CC. by 4.0): 156; Berlin Museum of Natural History, Germany: 193; Biblioteca dell'Accademia Nazionale dei Lincei e Corsiniana, Florence: 102 left; Biblioteca Nazionale Centrale, Florence: 41; Bridgeman Images: 6 (detail; The British Library, London); 59 (© Royal Institution); 139 bottom (© Veneranda Biblioteca Ambrosiana/Metis e Mida Informatica/Mondadori Portfolio); 144 (© Courtesy of the Warden and Scholars of New College, Oxford); 147 bottom (© Christie's Images); British Library, Oriental Manuscripts, Or 2784, courtesy Qatar Digital Library: 139 (top); Caltech Archives and Special Collections, Courtesy of: 113, 125, 127, 201; Caltech/MIT/LIGO Lab: 28, 29; Cambridge University Library, Department of Archives and Modern Manuscripts: 187, 222; 224 (Reproduced with permission of the Syndics of Cambridge University Library and William Huxley Darwin); © 2005 CERN; Photo: Maximilien Brice: 131; Cold Spring Harbor Laboratory Archives, Courtesy of: 195, 199; Deutsches Museum, Munich, Archive, CD87008: 160; Division of Medicine and Science, National Museum of American History, Smithsonian Institution: 163; ETH-Bibliothek Zürich, Rar 21896: 218; German Federal Archives, Koblenz (CC BY-SA 3.0 DE): 161; © Getty Images: 23 (Bettmann); 25 (Photo by Boyer/Roger Viollet); 49 (National Gallery, London/Photo12/Universal Images Group); 55, 108, 148 (Science & Society Picture Library); Getty Research Institute, Los Angeles: 36; Heidelberg University Library (PDM): 225; Institut Pasteur/Musée Pasteur, Paris: 77 (MP:30357); 78 (MP21012), 79; James St John, via Flickr: 95; © Jim Harrison Photography: 166; John Carter Brown Library, Providence: 138; Jolyon Troscianko, Courtesy of: 234; Library of Congress, Washington D.C.: 43, 44, 50, 63 (Rare Book and Special Collections Division); M. C. Escher's *Depth* © 2023 The M. C. Escher Company-The Netherlands. All rights reserved. www.mcescher.com: 99; Manchester Literary and Philosophical Society, The (Photo courtesy Science & Society Picture Library, London): 54; Marine Biological Laboratory Archives, Arizona State University, "Hans Spemann." *Embryo Project Encyclopedia* (1931). ISSN: 1940-5030 http://embryo.asu.edu/handle/10776/3167: 191; Metropolitan Museum of Art, New York: 70; Missouri Botanical Garden, Peter H. Raven Library, via BHL: 2 (bottom left), 105; Musée d'histoire des sciences de la Ville de Genève, MHS 2237 (CC BY-SA 3.0 FR): 117 (top); Museo Galileo, Firenze. Photo by Franca Principe: 40, 101; Museum Prinsenhof Delft, Collection. Gift of H.C. Vroom (photo Tom Haartsen): 35; Nadrian C. Seeman, New York University: 98; NASA: 37; National Institute of Standards and Technology Digital Archives, Gaithersburg, MD: 27; National Library of Medicine, Bethesda: 85, 209 (photo: Norman MacVicar), 210; National Museum Boerhaave, Leiden: 171 (photo Tom Haartsen), 229 (Niko Tinbergen's *A Total of 1095 Responses*); Natural History Museum Library, London: 2 (top left), 20, 68, 219; Nimitz Library, United States Naval Academy: 19; Northeastern University, Boston, MA, Snell Library: 107; Paul W. K. Rothemund, "Folding DNA to create nanoscale shapes and patterns," *Nature*, vol. 440,16, p.298, fig. 1 (detail), © 2006 Nature Publishing Group reproduced with permission of SNCSC: 97; PBA Galleries, Berkeley, CA, Courtesy of: 81; Philip Mynott: 233 (bottom); Photo: Kelvin Ma, via Wikipedia (CC0 1.0): 147 (top left); © Photo SCALA, Florence: 16 (CMN dist. Scala/Photo: Benjamin Gavaudo, 2016); 39 (Courtesy of the Ministero Beni e Att. Culturali e del Turismo); 189 (RMN-Grand Palais/Dist. Photo SCALA); PNAS; After Philip C. Hanawalt, "Density matters: The semiconservative replication of DNA," *PNAS*, Vol. 101, No. 52, fig.3. Copyright (2004) National Academy of Sciences, USA: 203; Private collection: 17; 192, 216, 230; Rijksmuseum, Amsterdam (CC0 1.0), SK-A-957: 170; Roslin Institute, The University of Edinburgh, Photo courtesy of The: 211; Royal Society, London, The: 32 (detail), 47, 122; Sarah Jelbert, Dr., University of Bristol: 233 (top); Science History Institute, Philadelphia: 2 (top right), 51 (top left); Science Museum Group: 103; Science Photo Library, London: 66, 73 (Royal Institution of Great Britain); 72 (Sheila Terry); 93 (J. Bernholc et al, North Carolina State University); 117 bottom, 119 (IBM Research); 129 (LANL/Science Source); 152 (Royal Astronomical Society); 203 (Steve Gschmeissner); 207 (A. Barrington Brown, © Gonville & Caius College); 215 (Thomas Deerinck, NCMIR); Sean R. Garner and Lene Vestergaard Hau, *Coming Full Circle*: 167; Shutterstock/pOrbital.com: 109; Smithsonian Institution Archives, Washington D.C.: 26, 53; Smithsonian's National Zoo, Photo: Roshan Patel: 221; Stanford University Libraries, Courtesy of the Department of Special Collections (M2568, Emil Fischer papers): 87; Suleymaniye Manuscript Library, Istanbul, by permission of the Presidency of Manuscripts Institution, Turkey: 137; SXS (Simulating eXtreme Spacetimes) Project, Courtesy The: 12 (detail): 30; © The Trustees of the British Museum: 67; University Libraries Leiden (CC BY 4.0): 34; University Library, Basel: 141; University of California Libraries: 15, 86; University of Cambridge, Courtesy of and Copyright Cavendish Laboratory: 111, 122 (Courtesy Royal Society, London); University of Illinois Urbana-Champaign: 22; University of Toronto, via BHL: 186 (Gerstein Science Information Centre); 232 (Robarts Library); Wellcome Collection, London (PDM): 2 (bottom right), 7, 8, 9, 10, 45, 51, 69 (bottom), 74, 80 (detail), 82, 83, 89, 91, 102 (right), 134, 143, 146, 150, 168, 172, 173, 174, 177, 178, 179, 185, 220, 227; (CC. by 4.0): 57, 188, (David Gregory and Debbie Marshall), 223; (Copyright, by permission): 205, 206; Copyright (2023) Wiley. Used with permission from P. Lenard, "Über die lichtelektrische Wirkung," Annalen der Physik, 1902, vol. 8, issue 1: 156 (top); The Yorck Project (2002): 180.

Author credits

Experience had taught me that the editorial and design team at Quarto—in this case, Ruth Patrick, Anna Galkina, Martina Calvio, Sara Ayad Cave, and Allan Sommerville—would do a beautiful job with the production of this book, and my expectations could not have been more fully met. A book of this sort is indeed a team effort for which the writer tends unjustly to get all the credit—although, I suppose, also all the blame. I am grateful too to my editor at the University of Chicago Press, Karen Merikangas Darling, who once again has placed in me a trust that I can only hope is warranted. For fielding a few queries about experiments in modern biology I thank Matthew Cobb, and I am grateful to eagle-eyed historian Thony Christie for casting a careful eye over the early chapters. I have benefited from discussions on the aesthetics of science with Milena Ivanova.